A
Field Guide
to
Economics
for CONSERVATIONISTS

A
Field Guide
to
Economics
for CONSERVATIONISTS

Brendan Fisher

Rubenstein School of Environment and Natural Resources,
University of Vermont
World Wildlife Fund

Robin Naidoo

World Wildlife Fund
Institute for Resources, Environment and Sustainability,
University of British Columbia

Taylor Ricketts

Gund Institute for Ecological Economics,
University of Vermont

Roberts and Company Publishers
Greenwood Village, Colorado

Library of Congress Cataloging-in-Publication Data

Fisher, Brendan.
A field guide to economics for conservationists / Brendan Fisher, Rubenstein School of Environment and Natural Resources, University of Vermont, World Wildlife Fund; Robin Naidoo, World Wildlife Fund, Institute for Resources, Environment and Sustainability, University of British Columbia; Taylor Ricketts, Gund Institute for Ecological Economics, University of Vermont; Roberts and Company, Greenwood Village, Colorado.

    pages cm
Includes bibliographical references and index.
ISBN 978-1-936221-50-9
1. Nature conservation—Economic aspects. 2. Ecosystem services. I. Naidoo, Robin. II. Ricketts, Taylor H. III. Title.
QH75.F55 2014
333.72—dc23

                 2014030208

Manufactured in the United States of America

10   9   8   7   6   5   4   3   2   1

Roberts and Company Publishers, Inc.
4950 South Yosemite Street, F2 #197
Greenwood Village, CO 80111 USA
Tel: (303) 221-3325
Fax: (303) 221-3326
Email: info@roberts-publishers.com
Internet: www.roberts-publishers.com

Publisher, Ben Roberts; production editor, Julianna Scott Fein; manuscript editor, Laura Kenney; creative director, cover designer, and illustrator, Emiko-Rose Paul; text designer, Linda M. Robertson; proofreader, Jennifer McClain. The text was set in 10.5/13.5 Utopia Standard Regular by Danielle Foster and printed on 60# Recycled White Opaque by Edwards Brothers.

Cover image: Fuelwood Collection in the Forests of the Massif de la Selle in Southern Haiti © Robin Moore.

About the cover: Forests in Haiti, such as those of the Massif de la Selle, harbor significant biodiversity and provide Haitians with a suite of benefits. They supply fuelwood for cooking and heating. They retain sediment during storms. They regulate water flows. However, the fate of Haiti's forests, like many forests around the world, is driven largely by economic decisions—do the benefits of conserving forests outweigh the benefits of chopping them down? This is just one example of why conservationists need a field guide to economics.

*This book is dedicated to conservationists around the world
working on the complex issues we introduce simply here.*

Contents

3 The Economist's Punch Line: Supply and Demand   39

4 Ecosystem Services: Where Nature Meets Human Welfare   63

5 Valuation Techniques: How to Estimate Nature's Benefits   85

Preface

Why do conservationists need a field guide to economics on their shelves alongside the well-loved bird and plant guides? Two reasons, really. First, the economic decisions people make every day are at the core of the world's conservation issues: climate change, Amazonian deforestation, tiger poaching, vulture declines in Asia, and countless others. Second, and more importantly, an understanding of the economic forces behind these decisions can help conservationists safeguard biodiversity in a more sophisticated and effective way.

That is why we wrote this book. We wrote it as a primer for conservationists of all stripes—scientists, activists, staff members of nonprofits and government agencies—who want to understand and use economic concepts in their work. And we wrote it for college and graduate students to use in courses or on their own to begin building an economics toolkit.

Why don't more conservationists already understand and appreciate economics? One reason is that, like most normal people, they've found the subject dreary and boring. So we've done our best to be different, aiming for a book that is straightforward, fun, and even irreverent at times. We also make no

attempt to be exhaustive, focusing instead on a small number of core concepts that we felt would be most useful to the most people. Our apologies to economists who think additional concepts should have made the cut.

You'll find that the book builds from chapter to chapter, starting with simple, fundamental concepts and then expanding from there. To illustrate ideas clearly, we establish a mythical landscape that grows more complex throughout the book. We complement these simple illustrations with examples from the scientific literature and from real-world conservation stories.

The fact is, economics is really cool. It explains a lot about the problems we face, and it will help us solve them. That is why we think everyone involved in conservation should know the basics.

Acknowledgments

We are grateful to many who helped in various ways along this book's journey from half-baked idea to the final "field guide" you hold in your hands. We thank the friends and colleagues who helped us field-test parts of the book's content in various public houses and classrooms, which helped us flesh out some of the examples in the text to make the economic concepts come alive—in particular, Andrew Balmford, Richard Bradbury, David Edwards, and David Wilcove. We thank our reviewers who pushed us to deliver better content and saved us from publishing some of our worst jokes. This bright and generous crew includes Elena Bennett, Josh Goldstein, David Hadley, Emily McKenzie, Jouni Paavola, Belinda Reyers, Brian Robinson, and Kerry Turner.

Thanks also to an inspiring group of scientists who have shown us all how important economics is to our most pressing conservation and sustainability issues. These leaders include Vic Adamowicz, Paul Armsworth, Andrew Balmford, Ian Bateman, Jim Boyd, Esteve Corbera, Gretchen Daily, Brett Day, Jon Erickson, Josh Farley, Paul Ferraro, Shuang Liu, Daniel Lopez Dias, Kenneth Mulder, Jouni Paavola, Subhrendu

Pattanayak, Derric Pennington, Alex Pfaff, Steve Polasky, Jim Sanchirico, Kerry Turner, and David Wilcove.

We thank Julianna Scott Fein, Emiko-Rose Paul, and Laura Kenney for taking a disjointed manuscript and our own less than stellar artwork and converting it into something readable and pretty. Many thanks to our publisher, Ben Roberts, for his continual support, good humor, and astonishing patience in putting up with us (particularly Robin) throughout this project. We are grateful to Bret Michaels, chickens, Greenbank Farm, the Dungeness crab of Puget Sound, and the English Premiership for keeping us fed and entertained during our writing get-togethers. And we are indebted to Small World Coffee, Stone Soup, Buzzin' Alawng, and New Moon Café for being the type of establishments that allow people to sit, write, goof off, and consume large quantities of coffee.

Finally, we need to thank our families and close friends, not because they contributed a darn thing to this project, but for putting up with us as we got together to write.

About the Authors

Brendan Fisher is a research associate professor at the Rubenstein School of Environment and Natural Resources at the University of Vermont. He spends much of his nonworking time playing hockey, soccer, and board games with his three children. Brendan's research focus is on the nexus of economics, ecosystem services, human behavior, and poverty alleviation. He is a fellow at the World Wildlife Fund, a fellow at the Gund Institute for Ecological Economics at the University of Vermont, and a fellow at the Centre for Social and Economic Research on the Global Environment (CSERGE) at the University of East Anglia. Brendan graduated the eighth grade from St. Joseph's School in Aston, Pennsylvania, with a solid B in social studies.

Robin Naidoo is Canadian and therefore gives this book a modicum of credibility. For the last decade he has worked as a conservation scientist for the World Wildlife Fund, investigating the ecology, economics, and conservation of biodiversity. He

works closely with the Community-Based Natural Resources Management Program in Namibia, where he gets to collar large and dangerous wildlife, to the chagrin of his office-based co-authors. He is an adjunct professor in the Institute of Resources, Environment and Sustainability at the University of British Columbia; fellow at the Centre for Social and Economic Research on the Global Environment (CSERGE) at the University of East Anglia; and an affiliate at the Gund Institute for Ecological Economics at the University of Vermont.

 Taylor Ricketts is professor and director of the Gund Institute for Ecological Economics at the University of Vermont. That makes him sound like an economist, but he really is a biologist who could have used this book to avoid a decade of trying to understand his coauthors. His research focuses on the overarching issue, How do we meet the needs of people and nature in an increasingly crowded, changing world? Specific work includes estimating the economic benefits provided to people by forests, wetlands, reefs, and other natural areas. In addition to his work at the Gund Institute, Taylor is a senior fellow at World Wildlife Fund. He considers the bees he studies to be as impressive as—and easier to collar than—Namibian wildlife.

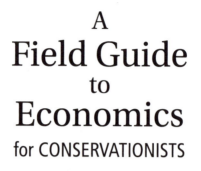

A
Field Guide
to
Economics

for CONSERVATIONISTS

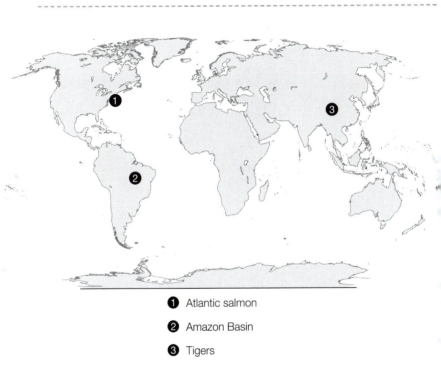

1 Atlantic salmon

2 Amazon Basin

3 Tigers

Introduction

Why Economics Is at the Core of Conservation

Perhaps no species more symbolizes wild nature than the tiger. Just 100 years ago roughly 100,000 tigers roamed across Asia, but since then hunting and poaching, human encroachment, and habitat loss have combined to reduce this number to about 3,500, and three of the nine original subspecies have gone extinct. In 2010, the Chinese year of the tiger, heads of all tiger-range states signed an accord to reverse this trend and double the number of wild tigers by 2022, the next year of the tiger. Will it work?

The Amazon Basin contains around one-half of all rainforests on Earth and houses an estimated one-third of the Earth's species. The area is so vast and unexplored that five species of primates have been discovered over the last decade. Despite its obvious importance and heroic conservation efforts, the Amazon lost about 20,000 square kilometers of forest a year between 1996 and 2005 (Regalado 2010)—about the size of

Wales, or 143 million average-sized Starbucks cafés.[1] In recent years, though, the deforestation has slowed; in 2010 the rate of loss was 80% lower than the 1996–2005 average. Why did this slowdown occur?

Once considered too abundant to sell, Atlantic salmon spawned in every major river from Long Island to northern Quebec in North America and from Portugal to Norway in Europe. By 2006 there were only around 1,200 wild Atlantic salmon spawning in US rivers—all of which were located in the state of Maine—and declines were similar in Canada and Europe (Wilcove 2008). In 1997 the United States responded by banning salmon fishing within its coastal oceans (exclusive economic zone, or EEZ), and Greenland and Canada have enacted similar legislation. Can the salmon recover?

Conservation Challenges and Economics

These stories are all—literally—textbook examples in conservation: one-time abundance, alarming declines, and a race to understand why and what to do about it. The stories certainly involve ecology, hydrology, biogeography, oceanography, politics, sociology, and ecosystem science.

But they hinge equally on economics. There . . . we said it. And when we say *economics* we're not talking about how the stock market is doing. Economics is much broader than that. Economics is about the broad world of understanding how the incentives we face affect the decisions we make.

---

[1] The average Starbucks is around 1,500 square feet. No, there are not this many Starbucks on the planet . . . yet.

We define economics more fully a little later, but trust us for now: the causes of biodiversity loss are grounded in economics. The land- or resource-use decisions that affect the natural world are grounded in economics. And most importantly, the solutions and their effectiveness are grounded in economics. Look again.

Various tiger body parts are prized for use in traditional Chinese medicine as cures to ailments like epilepsy, malaria, and sexual dysfunction. The extraordinarily high price that people are willing to pay for tiger penises and other body parts drives poachers to supply these goods to markets despite the risks and costs associated with the illegal hunting of tigers. Tiger habitat is also continually lost to make way for economic activities: mangroves are contracting due to expanded fisheries and agriculture, and forests in Southeast Asia are being turned into timber and oil palm plantations. And a burgeoning population hunts deer and other tiger prey for food of their own, while the high costs of patrolling protected areas make it easier for poaching to happen. If this trend is to be reversed, economic issues, such as the incentive to harvest and sell tiger parts, the profitability of forest conversion versus forest conservation, and the development of less damaging supply chains for palm oil and timber, will need to be addressed.

In the Amazon, drivers of deforestation have changed over time, but all are based on the pursuit of financial gains from land clearance. Logging operations not only result in cash bonanzas from the sale of timber, but the associated new roads open up land for cattle ranching, a major driver of land-use change in the region. More recently, global demand for soy has led to conversion of huge areas of pasture to soybean fields, displacing ranchers further into the forest.

So why, then, have deforestation rates declined recently? Again, the answer is wrapped up in economics. Some of the decline can be attributed to the global economic slowdown (Assunção, e Gandour, and Rocha 2012); there is simply less capital sloshing around to fund the timber-cattle-soy triumvirate. New analyses suggest that roughly 50% of the decline can be attributed to a set of policies that increase monitoring of deforestation activities, increase the area under protected and indigenous control, and restrict credit access to borrowers who can prove they comply with the forest code. All of these policies have an economic aspect. Monitoring costs money, but it also raises funds via fines. Parks and indigenous reserves increase the costs for those who would like to convert other forests to pasture or soy fields. Why is this? Competition. If there is less forest available to convert because of the protected areas, there are fewer places available to convert. Finally, the Brazilian government's decision to remove credit and subsidies for ranching and agricultural expansion, along with falling global commodity prices for agricultural goods, have combined to make the conversion of forests less profitable.

The reasons for the decline of Atlantic salmon are clear and numerous: overharvesting of the adult salmon at sea, dams and fishing in spawning rivers, agricultural pesticides and sediment in spawning rivers that poison or smother smolt, and diseases transmitted by pen-raised salmon. Many of these practices are the result of rational economic decisions by individuals seeking to maximize their own financial benefits. But these decisions have an effect on others, too, and can be detrimental to society as a whole when the larger, social costs and benefits of agriculture or aquaculture are not properly taken into account. Salmon recovery will depend on revising the economic signals

that have led to their decline; i.e., changing the incentives to act in ways that negatively affect salmon. Actions that might help wild salmon populations, such as reducing sediment erosion in spawning watersheds, limiting freshwater abstraction for irrigation, and mitigating pesticide use on such crops as blueberries, are often not taken or enforced because of the financial burden they levy on politically powerful actors, like logging companies or agricultural producers.

These cases—the decline of the tiger, the deforestation of the Amazon, and the devastation of the Atlantic salmon populations—are often trotted out to illustrate principles of population viability, trophic dynamics, landscape ecology, and reserve design. But they also illustrate principles of opportunity cost, externalities, cost-benefit analysis, and other central concepts in economics (which we'll explain later). And these cases are far from unusual. From the logging of the wintering grounds of the monarch butterfly to an anti-inflammatory drug given to cattle that has decimated Asiatic vulture populations,[2] from the invasion of eastern Australia by prickly pear to the global moratorium on whaling ... economics is often at the core of conservation issues and is typically central to their solution.

In a nutshell, that is why we have written this book. We believe conservation scientists should strive to be as adept with the fundamentals of economics as they are with such disciplines as ecology and natural history. We feel it will help them understand conservation problems more fully and devise solutions more cleverly. We think that paying attention to the economic aspects of conservation issues will increase the likelihood that solutions to conservation problems can be

[2] See http://www.rspb.org.uk/supporting/campaigns/vultures/

implemented. Of course, other social science disciplines, like anthropology, sociology, and political science, play key roles in improving conservation outcomes. We have no excuse for our scant coverage of these disciplines in these pages. We sat down to write a primer on economics; we made choices based on the scarcity of our time, our knowledge, and our patience with each other. What? Scarcity? See next section.

What Economics Is (and Isn't)

A common misunderstanding among ecologists and conservationists is the belief that economics equals moneymaking, and that economists therefore are inherently unsympathetic to environmental concerns and so are not to be trusted. In this view, economics is equated with markets, or gross national product, or simply getting rich at the expense of the environment. But economics is actually very different from any of these things. It is a scientific discipline, akin to ecology (with which it shares the Greek root *eco,* meaning "house"), rather than a doctrine like capitalism. People also often confuse economics with finance (the study and management of money), commerce (the buying and selling of goods), or business (the actions involved in making money/profits). Economics does cover some aspects of each of these but is much broader than any of them.

> **Economics** is the study of how people make choices under conditions of scarcity, and of the results of those choices for society (Frank and Bernanke 2003).

What *is* economics, then? At its most basic level, economics is "the study of how people make choices under conditions of scarcity, and of the results of those choices for society" (Frank and

Bernanke 2003). So, it's all about how people make choices and how people respond to incentives. What's with the "scarcity" bit? If money or time or resources weren't limited, we wouldn't have to make choices at all—we could do or buy everything we wanted. But given those limits, we have to evaluate trade-offs (Do I want new pants or a nice dinner? Should society subsidize health care or reduce national debt?) and then make difficult decisions.

Conservationists are certainly familiar with issues of scarcity. In the first instance, "scarce" or rare species are those that we are most concerned about conserving. We don't see a lot of commotion about the "Save the Mosquito Campaign" or the "Conserve Salt Water Coalition"; these things are abundant and are doing just fine without our attention. But this is not the exact scarcity we talk about in economics. In economics scarcity is a constraint that forces decisions. Conservationists are, sadly, all too familiar with the scarcity of financial and other resources at our disposal. We need to decide how we protect vulnerable populations of albatrosses or remove invasive species under scarce funding. We need to think about how to manage rare species in open-access conditions, like deep-sea fisheries, with limited information and management capacity. We need to design plans to manage our forests and grasslands so that we protect the greatest number of species with a limited budget.

In addition to economics being a discipline concerned with the allocation of scarce resources, economics is also a way of thinking. That might seem lofty and vague, but really, the most important thing you could learn about economics is what an economic approach to a problem looks like. What is the economic way of thinking about an issue? Well, picture this:

An economist and a biologist walk into a bar, sit down, and order a few drinks. If the biologist is Taylor, he'll no doubt order

a microbrew with some ironic name, and if the economist is Brendan, he'll order a "yellow" beer. They sit down and the biologist says to the economist:

> BIOLOGIST: *Hey, I'm headed to survey birds in Peru. Do you want to come?*
>
> ECONOMIST: *Well, what does that involve?*
>
> BIOLOGIST: *Hike three days in, eat only granola and beans, bury our own waste, stand silently listening for birds for seven days, and then hike out.*

The economist grabs a napkin and starts a list of costs of going along on the trip. He then thinks about the obvious benefits and lists them, too. The napkin then probably looks like a messier version of **Figure 1.1**.

The economist then looks at the two sides of the napkin and evaluates them in some way (e.g., counting up the number

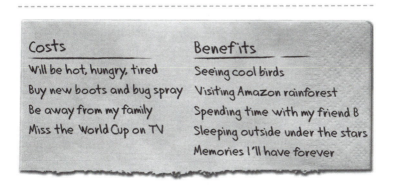

FIGURE 1.1 Cost and benefit table of an economist's decision to survey birds in Peru. (NB: The cost and benefit table of the biologist taking an economist to survey birds is likely to look very different.)

of costs and benefits, estimating the value of each, and then comparing the totals). In this case, it seems a good deal, so:

ECONOMIST: *I'm in!*

This situation is just a normal decision-making process of weighing the positives and negatives of a choice. In economics this process is formalized in cost-benefit analysis, but in general it simply means that most people will make a decision in which they do the best they can to come out ahead—that is, to maximize their net benefits. *Net* here means the benefits minus the costs. For example, the economist had to think about what he was giving up in order to go see the Amazon, such as time with family, the soccer games, etc.; then he had to weigh these costs against the benefits. This is the simplest version of an economic approach to decision-making. It will get more complex throughout the book.

Why You Need a Book about Economics

As you've just seen, economics is the science of making decisions on how to allocate scarce resources, and conservation also faces difficult decisions, under scarcity, every day. What agricultural benefits do we lose if we decide to conserve one more hectare of the Amazon? How do these lost benefits compare to what we gain from conservation? Is one more tiger in the wild worth the cost of protecting it? Who benefits from imposing high stream-quality standards for salmon, and who pays the costs of implementing these standards? How do we design a network of protected areas such that we're getting maximum conservation return on our investment? How do we

evaluate how much people care about a given species or natural area; i.e., is it important enough to people that we should fight to save it?

You can already see the trade-offs and cost-benefit analyses underlying these classic conservation problems. After reading this book you will be able to recognize the range of economic concepts they embody: opportunity costs, marginal analysis, distribution and externality concerns, consumer surplus, and economic valuation.

We hope it's clear by now that economics is often at the core of conservation issues. We feel that an understanding of economics is essential to understanding conservation issues fully, to studying specific problems carefully, and to intervening to conserve biodiversity effectively. Economics is as much a part of conservation science as ecology and natural history. In our increasingly crowded, consumptive, and globalized world we think that the most effective conservation scientists will be those who are as adept with economics as they are with the more traditional natural sciences.

A Primer on Economics for Conservation

This book is intended to be a primer in the true sense of the word—a brief introductory text. In this primer, we attempt to lay out the economic principles that are most important and relevant to conservation. The book is meant for those interested in both the science and practice of conservation and who want a working understanding of economics for that purpose. It covers what our biologist author, Taylor, wishes he learned long ago about economics.

We kept our list of what the book *is* short so that we didn't have a lot to live up to. That leaves a long list of things this book *is not,* but we'll just highlight the top three. First, this book is certainly not an exhaustive treatment of every concept in economics that is pertinent to conservation issues. We have chosen the concepts that, in our experience, are at once most central to conservation and also most bedeviling to conservationists. Second, we mostly focus on concepts, not methods. For example, the design and analysis of particular techniques, like choice experiments, are introduced but not covered in any great detail; there are other books for that. Finally, this book is not big. It's only 185 pages—even Brendan could read it cover-to-cover in a week.

We organized the book to build up from the most fundamental principles and concepts to more nuanced ones, chapter by chapter. Woven through all the chapters is a simple example that grows as the book progresses and as we incorporate additional concepts. After illustrating new concepts in this simplified way, we then get messy, examining cases and analyses from the conservation literature and practitioners. Our aim in this two-pronged approach is to demonstrate both how economic concepts *can* inform conservation and how they *have* (or *haven't*) in real life. Throughout, we draw on the direct experiences of conservation practitioners to bring the concepts to life and show how economics is central to many of today's conservation issues.

In the next chapter we set out the critical economic concept of opportunity cost. This idea sits at the very core of economic decision making and therefore economics as a way of thinking. We also discuss the major economic tool for policy and project assessment, cost-benefit analysis. In Chapter 3 we look at the relationship between supply and demand and

demonstrate how we typically make decisions at the margin; i.e., we make decisions based on small changes relative to our current situation. This fact affects our assessments of trade-offs and therefore affects any formal or informal cost-benefit analysis. In Chapter 4 we discuss the public-private good framework that can be used to understand how the essential qualities of a good affect the way it can be supplied and managed. This helps us to understand why the market does a decent job supplying us with pants but not such a good job ensuring that the ozone layer doesn't get a big hole in it. Here the concept of ecosystem services comes most explicitly into the book.

Chapter 5 is largely about demonstrating the benefits of conservation in a quantitative way. We discuss valuation techniques and approaches as well as the different metrics with which conservation benefits can be measured. Demonstrating that conservation has benefits is one thing, but *capturing* these benefits in a way that is meaningful to actors making decisions that affect conservation is quite another. This is what Chapter 6 is all about . . . institutions. In economics *institutions* is really a broad term encompassing the rules and norms that describe how society works. Key ideas in this chapter include property rights, information failure, and market failure as well as mechanisms to capture benefits (e.g., payments for ecosystem services and certification schemes). In Chapter 7 we take all of the concepts built up in the text so far and utilize them in a semi-quantitative conservation planning example. This chapter builds on the simplified system used to explain concepts throughout the book in an attempt to tie together all of the pertinent methods and techniques that might be useful to readers in thinking about their own conservation work. In Chapter 8 we

jam all of the stuff that was too complex for earlier chapters, stuff we oversimplified, and briefly some things that we didn't cover but were simply too cool to leave out.

Although this book might not change your views on economists, we hope that it does help you to appreciate the usefulness of economics for conservation. Conservation science is inherently problem based. When faced with a problem you need all the tools appropriate to the job. Here we strongly advocate that economics is one of the tools needed to confront our current and future conservation challenges.

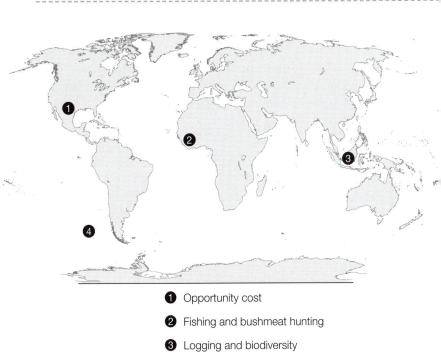

1. Opportunity cost

2. Fishing and bushmeat hunting

3. Logging and biodiversity

4. Cost-benefit analysis

Opportunity Cost and Cost-Benefit Analysis

Why Conservation Often Loses Out to Other Stuff

Imagine that you live near the coast in Ghana and spend a lot of time fishing. Most days you go out and try to catch enough fish to eat and perhaps to sell a few in the market. However, some days you do not go fishing, but instead you take to the forest and hunt animals for meat. This "wild meat" (or bushmeat, as we'll refer to it from here on) again could be for eating or selling. When you wake up on a given morning, how do you decide whether to go fishing or to go into the forest to hunt?

Of course, this decision may be affected by a dozen things: Are you too tired for fishing? Do you think there will be lots of fish and little wild game? Do you think there will be a lot of demand for fresh fish in the market today? The factors that go into making your decision are often economic in nature, but here we want to draw special attention to the fact that when you decide to go fishing for that day, you have given up the opportunity to go hunting, and vice versa. Some of the costs of deciding to go hunting are the benefits that you would have gained if you had gone fishing. Let's say that again, and read it slowly: *Some*

of the costs of deciding to go hunting are the benefits that you would have gained if you had gone fishing. This is the concept of an opportunity cost.

For example, if on an average day of fishing you catch 60 kilograms of fish, your benefit is 60 kilograms per day. If on an average hunting day you get 20 kilograms of bushmeat, your benefit is 20 kilograms per day. However, the cost of going fishing is forgoing the 20 kilograms you would have acquired if you had gone into the forest to hunt. Therefore, the *net* benefit (meaning the benefits minus the costs, measured in kilograms here) of going fishing instead of hunting is 40 kilograms per day (of course there are other costs, but we'll get to those later).

> The term **net** means that costs have been accounted for in an assessment of the benefits, and vice versa. Net simply indicates the benefits minus the costs; e.g., your net income is total money earned (gross income) minus costs, such as taxes.

In economics-speak, the opportunity cost of going fishing is the 20 kilograms of bushmeat you gave up by not going hunting. The concept of opportunity cost is described in probably the most-used economics textbook in the world (Mankiw 2003) as "the cost of something is what you give up to get it." A slightly more technical definition of opportunity cost is the net benefits forgone from the next-best alternative that was not pursued.

That's a mouthful, but the concept of an opportunity cost is fundamental to economic decision making in any realm, including conservation. Every decision you make means that you cannot do something else. For example, in the last chapter Brendan was deciding whether or not to go survey birds in Peru with Taylor. Robin now joins them at the pub and tries to

order a beer. The bartender says, "There's a one-beer minimum here and we only have Budweiser and Guinness. They cost the same." Robin says, "Guinness, please." What has the Budweiser become in this situation with a one-beer minimum? An opportunity cost. Robin has forgone the Budweiser (his next-best alternative) for a Guinness. So why did Robin choose Guinness if it costs the same as Budweiser? It's because of another, sometimes awkward economic term—

utility. Robin gets more utility from Guinness than he does from Budweiser. Utility is simply the level of satisfaction someone gets from a good, service, or activity. If it sounds pretty general and subjective, that's because it

> **Utility** is another general term used in economics that at times is equated with welfare, satisfaction, or even happiness (Pearce 1981). In short, utility is the satisfaction one derives from a good, service, or activity.

is. It's a term that tries to wrap up all of the tangible and intangible benefits of a decision into one word. So in economic terms Robin gets more utility from Guinness, or in the normal world . . . he likes Guinness better. So while Robin is soaking up his utility, Brendan says, "Hey Robin, I'm heading to Peru next month to survey birds." Robin says, "You don't know a damn thing about birds." He continues, "It's a real shame too, because I was going to invite you to a World Cup qualifying game." If Brendan really wants to do both, but he sticks with the Peru trip, then the World Cup game becomes an opportunity cost of going to Peru.

You can now see how the choice of one activity always causes a person to forgo another, even if that other activity is "doing nothing." Remember from the last chapter we defined economics as "the study of how people make choices under conditions

of scarcity, and how these choices affect society" (Frank and Bernanke 2003). In the hunting-fishing example, you have to make a decision between the two activities. You have to make a choice under a condition of scarcity (your time) and that choice affects society. In this case, the choice affects society in terms of your own private benefit from the day's activity, the welfare of your family, or the ability of people in the marketplace to purchase fresh fish.

Crucially, your choice also affects society in a public or social way, because it has an impact on the reefs and forests you decide to exploit. The choices of fishers and hunters determine the pressure on wild populations of parrot fish, sharks, baboons, porcupines, antelope, and all other species that people harvest for food. This is one of the fundamental ways that opportunity costs help us understand the economic forces that are threatening biodiversity.

You might be saying to yourself, why would I go hunting instead of fishing on any given day, if I always get three times the amount of meat from fishing? Well, if a kilogram of meat was just a kilogram of meat, then it would always make sense to go fishing. But we know this to be untrue. We all have preferences for some foods over others. Assuming all else is equal (which it most assuredly isn't), preferred foods will fetch higher prices. We'll get to the nuts and bolts of this principle (supply and demand) in Chapter 3. For now, though, let's say in our market in Ghana, bushmeat commands twice the price per kilogram compared to fish—say, USD $2 instead of $1. Now, for a full day's work in the forest you can earn $40 by selling bushmeat in the marketplace and $60 by selling fish. It would still make sense to go fishing; the opportunity cost of going fishing is now

$40 (again, the forgone benefit of the alternative). But your net gain is still $20 ($60 – $40). However, when the price of bushmeat hits $4 per kilogram, the opportunity cost of going fishing hits $80, and you realize that by choosing to fish you have now lost $20 in income for that day ($60 you get from fishing minus the $80 you gave up by not hunting).

Certainly in the real world, things are much more complex. First, you cannot guarantee that you will have an "average" day of fishing or an "average" day in the forest. Second, people do not eat on "average." They need to eat pretty much every day, so deciding what activity you do for that day can have profound implications for your welfare. And third, this example assumes that you sell all that you harvest. If you are eating some of your catch, you will factor the value of your own consumption of meat into these calculations of opportunity costs.

Opportunity Costs in Ghana: Fishing and Bushmeat

Our fishing-hunting example was inspired by work that Justin Brashares and colleagues (2004) carried out in Ghana. Hunting for wild meat is widely considered one of the primary threats to biodiversity throughout the forested regions of Africa (Fa et al. 2002). Even if forests are protected on paper and look pristine from the air, hunting for wild meat often results in "empty forests" that are devoid of the larger and more conspicuous mammal and bird species. Figuring out what drives hunting for wild meat is therefore an important conservation question (Rowcliffe et al. 2005, Fa et al. 2009, Kumpel et al. 2009).

Brashares et al. looked at 30 years of data on fish catches and mammal populations in forests and found that fish and wild meat act as *economic substitutes*. In years when the supply of fish was poor, there was increased bushmeat hunting and therefore sizable declines in mammals in Ghanaian nature reserves. The data from the Ghana research looks a bit like **Figure 2.1**. For the years 1999–2003 there was a clear negative relationship between the amount of fish supplied per month and the amount of bushmeat supplied per month. Looking at Figure 2.1 we could imagine that when those supplying the market chose to hunt, they could not also go fishing (dashed area A). When they chose to go fishing they could not also choose to hunt (dashed area B). That's a classic trade-off in terms of how the suppliers spend their time—the opportunity cost of going fishing is the lost benefit from hunting, and vice versa.

Hal Varian (big-shot economist, writer of key economics textbooks, and now the chief economist at Google) says that "if the demand for good 1 goes up when the price of good 2 goes up, then we say that good 1 is an **economic substitute** for good 2" (Varian 2009).

That's pretty cool already, but Brashares et al. went further. They showed that this pattern was driven by fluctuations in fish populations. When stocks were low, people went hunting instead. This suggests that between areas A and B in Figure 2.1, people were continually revising their estimates of opportunity cost and choosing to hunt or fish accordingly.

This study by Brashares and his friends has become a textbook example in conservation science. Read through the paper and you won't find the term *opportunity cost* mentioned once, though the paper is largely about opportunity costs and economic substitutes. Economics is all around us in conservation; it just goes unnoticed. But understanding how economic

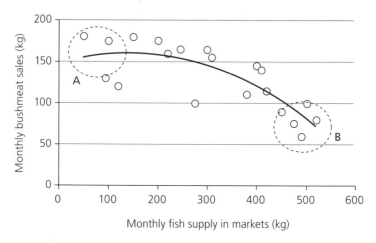

FIGURE 2.1 Relationship between bushmeat and fish sold in Ghanaian markets. (Simplified from Brashares et al. 2004.)

principles drive ecosystem change will help us do better conservation science and, ultimately, better conservation.

Previously we put dollar signs on the fish and bushmeat and assessed the opportunity cost in monetary terms. In Figure 2.1 we are looking at opportunity cost in terms of kilograms of bushmeat versus kilograms of fish. We saw with Robin choosing Guinness over Budweiser that opportunity cost wasn't just about the monetary price of an object, since both beers cost the same. Rather, Robin was thinking about what he was giving up if he chose the yellow beer instead of the black one. For Robin this was a no-brainer as the benefits (or utility) of Guinness far outweighed those of the yellow beer. The decision was about what delivered the most utility to Robin for a given cost.

But decision making in the real world isn't that easy; it's not all Guinness and yellow beer. Let's look at one more example of opportunity cost from the conservation literature where the two choices aren't even measured by the same metric (like kilograms versus kilograms or dollars versus dollars).

Opportunity Costs in Borneo: Birds and Logging

The lowland rainforests of Southeast Asia have been home to some of the most rapacious logging in the past half century. In some of the forests on the island of Borneo, loggers were pulling out more than 150 cubic meters of tropical hardwood per hectare in the 1970s, 1980s, and 1990s. What is 150 cubic meters? Well, it is almost four times the amount of wood removed in a typical logging operation in the Amazon. However, a lot of this logging has happened with two distinct cuts—sometimes separated by 30 years. What you are left with in these landscapes is three types of forests: primary forests, forests that have been logged once, and forests that have been logged a second time (i.e., going back in and harvesting all the smaller trees you missed before).

Imagine now that the Malaysian or Indonesian government is trying to make a (simplified) decision between more logging and more conservation in these lowland forests. Some information they would want to have is an understanding of both the timber values and the biodiversity values of the forests. On the biodiversity side, David Edwards and colleagues (2011) wondered how these different levels of timber extraction affected biodiversity. They measured birds and dung beetles in each type of forest: unlogged, once logged, and twice logged.

Somewhat surprisingly, they showed that over 75% of the bird and dung beetle species found in the unlogged forests were still present in the once-logged and even twice-logged forests.

On the timber value side, this group then looked into the financial costs and benefits of logging in these forests. They used logging data and cost estimates to calculate what profit could be made by logging such forests. It turns out that logging a primary lowland forest could deliver around $6,000 per hectare (net) in a first cut and another $2,000 per hectare (net) after a second cut.

With this information we can construct a trade-off curve that can graphically represent the opportunity cost of conservation in these forests. **Figure 2.2** shows this trade-off. We see that in an unlogged forest, the bird surveys of Edwards et al.

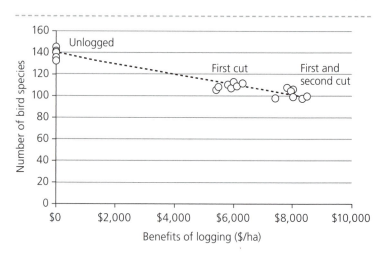

FIGURE 2.2 Relationship between the number of bird species found in forests in Sabah, Borneo, and the financial returns to different logging pressures. (Derived from Edwards et al. 2011 and Fisher et al. 2011.)

revealed that at least 140 species were present, but the monetary benefit of these unlogged forests for timber revenues is zero (although standing forests certainly provide other benefits). However, when we go in and log the forest, we can make around $8,000 (net) per hectare. But we lose about 40 species of birds from such a decision.

In this case, if the Malaysian or Indonesian government is trying to make the decision between two alternatives—logging these forests or conserving them for biodiversity—then the data we have here suggests that the decision to conserve the forest imposes an opportunity cost of perhaps $8,000 per hectare over the course of two cutting cycles. For this cost, we get to keep about 40 additional bird species, along with other benefits (wild meat hunting) and species that would go missing if we logged. Thinking of it the other way, the opportunity cost of logging is at minimum the loss of close to 60 bird species from the forest. Here, the costs and benefits are measured in different ways. An economic way of thinking can help us understand these trade-offs regardless of the way they are measured, but making the appropriate societal decision to conserve or cut a forest will require much more than an assessment of opportunity costs. (See **Box 2.1** for an example of how opportunity costs could be the driving factor in decisions to place protected areas across the globe.)

We will continue discussing opportunity costs throughout the book, because it can be a complicated concept to fully comprehend. For example, in a recent paper, over 75% of PhD-level economists failed to identify the opportunity cost of attending an Eric Clapton concert instead of a Bob Dylan concert (Ferraro and Taylor 2005). If you don't know who Dylan and Clapton are, then substitute them with Lady Gaga and Beyoncé.

BOX 2.1 Opportunity Costs and Protected Areas

The concept of an opportunity cost is nicely illustrated by looking at the placement of protected areas around the globe. Protected areas are a cornerstone of biodiversity conservation, so in principle one would want them to represent the range of habitats. It turns out, though, that they are not evenly (or even randomly) distributed across the globe, across nations, or across ecoregions (Joppa and Pfaff 2009). On average, protected areas are more likely to be found in high, steep locations far from cities and roads. Perhaps this is true because we've been busy protecting "pristine" areas, but another explanation comes to light when we think about opportunity costs. By siting protected areas in rugged, remote places, countries minimize the opportunity cost of conservation. They don't lose other opportunities to utilize lands for agriculture, forest extraction, or amusement parks.

A bunch of studies have shown these biases either globally or regionally. For example, Jon Hoekstra and colleagues (2005) evaluated protection levels among the world's 800+ ecoregions and found that those that were high, cold, or otherwise less fit for agriculture tended to have a larger amount of protected area. Going further, Joppa and Pfaff (2009) show that within countries protected areas with higher protection status tend to be located on higher and steeper terrain and in areas more remote than their less protected brothers and sisters. So protected areas closer to cities are more likely to be available for activities like timber production or farming, because there is a real cost of complete protection (i.e., forgone wood and food). Since the fully protected PAs are likely to be far, far away from urban demands, the opportunity cost is lower.

Cost-Benefit Analysis (A Brief Introduction)

So what do we do about this trade-off between saving birds in Borneo and logging these lowland rainforests? In the work of Edwards et al., none of the bird species that were lost from the landscape were endemic to the lowland forests in Borneo. Meaning, when we lose these species from the landscape they are not going extinct globally. Still, there are people who think that no amount of logging revenue is worth the loss of a single bird species to this landscape. On the other side of the debate, there are those who think no bird species is worth the opportunity cost of conservation (meaning, no bird is worth giving up the benefits of logging). And of course these extreme positions overshadow the likelihood that most people probably lie somewhere in between—they might be sympathetic to conservation, but they also understand the need for logging.

What is the answer in these cases where different members of society have different wants, needs, and opinions on a decision such as how to best use a forest? What if there are lots of opportunity costs and many different forms of utility all wrapped up into the same decision? The situation raises important questions of who benefits from a given decision, who pays the costs, and how we judge the significance of these trade-offs, these costs, and the benefits.

There are several disparate ways we can try to better inform the decision on what is the best use of the forest. There could be a public debate. People could propose a referendum and

hold a vote on what to do. Different *stakeholder* groups could petition the government to enact a piece of legislation to protect the forest for conservation or to open it up for logging.

A **stakeholder** is any person or organization that has an interest in the outcome of a decision. Stakeholders can be affected directly or indirectly by the result of a decision and therefore may want to take an active role in shaping the outcome of a decision-making process.

Another way to weigh the pros and cons of such a decision (and the one most applicable to our book) is cost-benefit analysis. In its casual form, cost-benefit analysis is just the simple comparison of the costs and benefits of a decision, like our trip-decision napkin in Chapter 1. In cases where the benefits are greater than the costs, the policy option under consideration is recommended. Or in the case of multiple options, the one where the benefits outweigh costs by the greatest amount. Again, picture Robin in the pub deciding which beer to choose, Guinness or Budweiser. Because he has made similar decisions before, he performs an almost automatic cost-benefit analysis, and the benefits of Guinness outweigh the costs of not choosing Budweiser. "Guinness, please."

This is cost-benefit analysis in its informal sense, but in the economics tradition cost-benefit analysis (CBA) has a long history, and it has become a more formalized process to aid in decision making in issues such as the forest problem discussed here. We will continue to learn about the process of undertaking CBA and how it applies to conservation issues throughout this book, but first we need to understand its history a bit.

A Truncated History of Cost-Benefit Analysis

The early theoretical foundation for CBA comes from the economic concept of Pareto efficiency. This idea is named after Vilfredo Pareto, probably the most famous economist with a rhyming name, and who like other famous economists (think Marx) had a great beard.[1] Pareto's work led to the idea that an *efficient* outcome (be it from a project, policy, or market) was one in which no one could be made better off without making someone else worse off. That's the ideal. You've squeezed all the benefits into a situation, such that any more benefit for someone always leads to a loss for someone else.

That's clearly a pretty tough standard to meet. In many conservation decisions, at least someone is likely to be worse off afterward. Think about the creation of a marine protected area (MPA). Sure, there are bound to be people who benefit (like vacationing snorkelers). But there are likely going to be people who are worse off (at least in the short run)—for example, the fishers who used to harvest exclusively inside the MPA. So Pareto efficiency is a hard goal to attain. Thank goodness for Kaldor and Hicks. Who? Nicholas Kaldor and John Hicks were the hottest indie rock band of the 1930s,[2] but they also did some economics on the side.

Both Kaldor and Hicks knew strict Pareto efficiency was nearly impossible to achieve, but they both spent time (separately) trying to make Pareto's ideas more operational. When you combine their two insights you get Kaldor-Hicks efficiency, which suggests that a project or policy is a "good" thing if

---

[1] He also is alleged to have lived his last days in a Swiss villa with 18 cats.

[2] This is a total lie; Kaldor and Hicks were martial artists. OK, that's a lie too.

the winners of the policy are, in theory, able to compensate the losers and still have some gain.[3] This theory is primarily about economic efficiency. So with our MPA example, could the winners (snorkelers, divers, etc.) in theory compensate the losers (displaced fishers)? Kaldor-Hicks efficiency has now allowed for the possibility that some people may be worse off, but as long as the net (remember: gains minus losses) is positive, then the project is a good idea.

This is where cost-benefit analysis comes in. In broad strokes, CBA compares the benefits and costs of (typically) a project or policy. The benefits are defined as those "things" that produce a gain in welfare. Conversely, the costs are the "things" that deliver a loss of welfare. CBA aggregates these gains and losses for all stakeholders in the decision, thereby conveying the idea that these are net social gains or losses (Pearce 1983).

We casually slipped the word *social* in there, but it's hugely important to this book and to conservation, so let's stop and consider it a bit. Economists use *social* in counterpoint to *private*. A private gain is one in which you are the only beneficiary. So when a landowner cuts her trees and sells the timber, her gain is a private gain. However, when that same owner plants trees along a river to help mitigate sedimentation, the gains here are social. Many people may benefit from enjoying a vegetated riverbank, shade from the trees, and better water quality. Costs can similarly be private or social. When a storm destroys someone's house, the cost is private. When someone dumps

---

[3] In truth, that definition is the Kaldor side of the issue. The Hicks side actually represents the situation when the losers are willing to compensate the gainers of a decision enough so that the gainers actually forgo the gain. But that is a pretty cumbersome thing to say in the main text.

trash into a river and affects the recreational enjoyment of local users or water quality, these costs are social. OK, back to cost-benefit analysis.

In formal CBA the changes in welfare are actually measured by an individual's willingness to pay (WTP) for the gain and an individual's willingness to accept (WTA) compensation to incur a loss. Huh? Let's unpack that a bit. Consider again the prospect of instituting a marine protected area. If you are a tourism operator, such a project may bring in more business, as tourists like to go to protected areas. Perhaps the snorkeling or scuba diving is better with an MPA than without. You might be willing to pay a certain amount of money to see that a proposed MPA actually becomes an official park. But you might also be willing to pay a certain amount for something you don't actually possess or have access to, simply because it is important to you. (Have you ever seen a polar bear? Would you contribute to a campaign to protect their habitat?)

On the other hand, if you are a fisher who uses the area slated to become the protected area, you might be willing to accept a payment or other compensation to offset the losses you anticipate. This is a WTA for a loss. In contrast to WTP, WTA is the correct measure of welfare when you already have rights to a particular good, service, or area. Economists have noted in hypothetical surveys that WTA measures are often a lot higher than WTP measures, possibly because once people have access to something, they can be very unwilling to give it up, especially in instances they consider less than fair. So people's WTA for the sale of poor-quality agricultural land in a forest enclave might be higher than someone's WTP for the same plots, because such issues as the history of residence and non-monetary connection to the area factor into their assessment of its value. More

parochially, you can think of a sports or theater fan who might pay $50 for a ticket to a game or show, but once they have it they wouldn't sell it for less than $100. Again, this shows the difference between the cost of a ticket ($50) and the utility derived from it (>$100).

Introducing Arden

OK, let's mock up a cartoon example and see how we might be able to apply some of our concepts to a conservation problem. Throughout the book we'll be building up this example, increasing its complexity to more closely align it to the complexity found in the real world.

For now imagine a simple landscape conveniently made up of six parcels of equal area; let's call this landscape Arden[4] (**Figure 2.3** on p. 32). The top three parcels (outlined in gray) are owned by Isabella.[5] She is a farmer thinking about converting her forest parcels into cropland. She has recently realized that, by keeping them forested, she is incurring an opportunity cost (i.e., the income she could get by farming them) that she doesn't want to bear anymore. So she crunches the numbers and realizes that she can get $500 for the timber that she removes from two

---

[4] Arden is the fictitious forest in Shakespeare's *As You Like It*. In this play the character Duke Senior, speaking of life in the forest, says, "And this our life, exempt from public haunt, finds tongues in trees, books in the running brooks, sermons in stones, and good in everything." Fitting words for conservationists?

[5] In the past decade, the name *Isabella* or *Isabelle* has been in the top 10 most popular female names in Aruba, Australia, Brazil, Canada, Denmark, England, Iceland, New Zealand, and the United States. So we thought it a fitting name for our Arden protagonist.

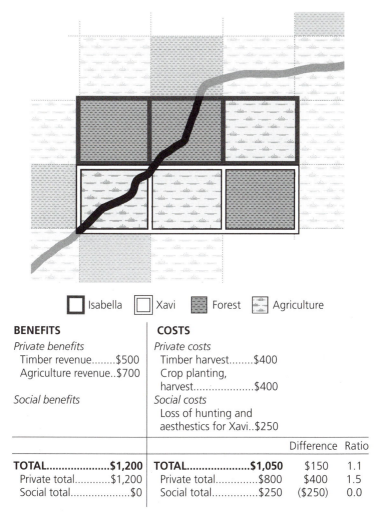

| | Isabella | | Xavi | | Forest | | Agriculture |

BENEFITS

Private benefits
  Timber revenue........$500
  Agriculture revenue..$700

Social benefits

COSTS

Private costs
  Timber harvest........$400
  Crop planting,
  harvest...................$400

Social costs
  Loss of hunting and
  aesthestics for Xavi..$250

		Difference	Ratio
TOTAL..................$1,200	**TOTAL....................$1,050**	$150	1.1
Private total...........$1,200	Private total..............$800	$400	1.5
Social total....................$0	Social total...............$250	($250)	0.0

FIGURE 2.3 Stylized landscape of Arden, with numbers on the costs and benefits of Isabella converting her forest to agriculture.

parcels. She also knows that she can get $700 for the crop that she will grow over the next year on this piece of land. If you add the two together you get a gross revenue of $1,200.

Is this $1,200 the entire opportunity cost of conserving this parcel as forest? No. Remember the opportunity cost is a "net" value. Isabella estimates that it will cost her $400 to cut down all the trees and remove the stumps prior to planting. She also knows that planting, weeding, and harvesting her crop will cost about $400 over the year. Now we can see that her net profit from conversion is $400 (Figure 2.3). This now is the opportunity cost of keeping this parcel forested. These costs and benefits to Isabella are all private, but remember that CBA is designed to take into account social as well as private costs and benefits for various stakeholders.

Isabella's neighbor Xavi[6] owns the parcels just south of hers (outlined with double lines in Figure 2.3). He hears that Isabella is thinking about converting her forest parcels. Several times each year, Xavi is able to hunt animals that come out of her forest onto his land. This provides Xavi and his family with a considerable amount of meat over the year. He also thinks that there are other neighbors who benefit from hunting wild animals, and he knows that though converting these two parcels might not bring an end to his hunting, it might cause a decrease in the number of animals available. From the viewpoint of the neighbors, Isabella's decision to cut down the forest incurs a social cost via the loss of available wild meat. Xavi estimates this loss to be worth about $200 for the year—that is, if he had to replace the lost protein by buying meat in the market, he'd

---

[6] Xavi is not a popular name as far as we know, but he is an amazing footballer.

have to spend about $200. However, Xavi considers his WTP for those parcels to remain forested to be $250. This extra $50 reflects other values that Xavi may have for the forest that are not related to wild meat, like his joy in hunting or the other foods or medicines he harvests from the forest, or simply his intrinsically positive feelings toward forests.

If we add up the costs and benefits of both stakeholders, we can see that the benefits of converting the parcel of forest are greater than the costs incurred by converting it (Figure 2.3). The benefits equal $1,200. The costs are $1,050. The net benefit is $150 for this parcel over the next year. Often in CBA analysis we report a benefit-to-cost ratio. In this case the benefit-to-cost ratio is $1,200/$1,050, or 1.1. As you can see, if the ratio is greater than 1, then the benefits outweigh the costs. If the ratio is less than 1, then the costs outweigh the benefits.

OK, if Xavi and Isabella have really taken into account all of the relevant costs and benefits, we can see that the conversion of the forest to crop makes sense. The net result is a gain. At the same time we can see that even though the net result is a gain, Xavi still loses out. Let's check in with Kaldor and Hicks—could Isabella compensate Xavi's loss and still make out? Check! Isabella could pay Xavi $251 for his losses and still make a profit by converting the parcel. You can see that the CBA already internally took into account the Kaldor-Hicks criterion. Of course, just because Isabella *could* compensate Xavi doesn't mean she *would* compensate him, but nonetheless, according to the Kaldor-Hicks criterion Isabella's decision is a net benefit to society.

If we imagine that Xavi's WTP was $1,000, the conversion of the land would fail the cost-benefit test. Sure, Xavi could compensate Isabella $401 not to convert the forest, and everyone

wins. But with a WTP of $1,000 the cost-benefit test fails whether or not Xavi compensates Isabella at all! This is a critical aspect of Kaldor-Hicks efficiency and of CBA—that compensation does not actually have to occur. CBA is an attempt to measure the net *economic* value of a decision and is not a technique to ensure justice or fairness or compensation. The compensation question is more akin to a *financial* value. In the next chapter we will examine the difference between the two values and their importance in shedding light on conservation questions, but you can imagine that Isabella's decision is unlikely to change based on Xavi's *theoretical* WTP to keep her land forested. CBA can help to inform us on what decision might yield the best social outcome, but actually implementing that outcome might require further agreements, contracts, policies, or markets—something we'll discuss in Chapter 6.

Our current Arden example is a simple one. There are only two people in "society" here, but what happens when there are other stakeholders to consider? We have also ignored a lot of costs and benefits in this example, things like the potential value of pollination services and the purification of drinking water that this standing forest might also provide. We have also not touched on issues related to the timing of these costs and benefits: What about the returns from farming after this first year? What if the price of the crop falls by 50%? What happens if it doubles in price? All of these things are going to change the CBA and the decisions on the ground, and they are all issues we will get to in subsequent chapters (but see **Box 2.2** on p. 36 for a bit more on the history and uses of CBA). The best launching point for discussing these issues is by first developing an understanding of the economic concepts of supply and demand—the focus of Chapter 3.

BOX 2.2 Cost-Benefit Analysis, Warts and All

Cost-benefit analysis (CBA) has been a popular form of project and policy evaluation, particularly in the United States, since it was first used to evaluate the merits of water projects in the 1930s. In fact, federal agencies in the United States are currently required to conduct a CBA for all significant regulatory actions, including sulfur dioxide emission from coal power plants and changes in drinking water standards. This process is also used in other countries and contexts, such as landfill taxes and policies in the United Kingdom or with regard to health warnings on tobacco products in Australia.

Remember that CBA simply sums up changes in welfare (i.e., costs and benefits) in order to calculate whether the net change of a decision is positive for society. The major critiques revolve around problems with deciding on the pertinent costs and benefits for a project, discounting, and equity and distribution issues. For example, can we actually calculate all of the costs and benefits of a decision to convert primary rainforest in the Congo Basin? Do we know what we are losing and how our welfare would change after its loss? The problems related to discounting we will come to in later chapters, but discounting is all about how we value the future, which is a difficult thing to do.

So, given the complexity of the interactions between our social and ecological systems, do we even know how to value future costs and benefits? For example, the potential loss of Arctic ice and polar bear habitat in the Northern Hemisphere

after 2100 should factor into a CBA regarding how to mitigate climate change, but can we actually estimate those future effects in a sensible way? And of course there are distribution and equity issues. CBA aggregates costs and benefits to individuals without regard to their identity. The important questions of *who wins* and *who loses* are ignored in the most basic form of CBA. Depending on this distribution, a CBA might recommend going ahead with a project that entrenches existing inequities. On the other hand, conservation can sometimes come at the expense of long-term economic development in poor rural areas. This is a classic problem in conservation. For example, there might be large private benefits of an industrial enterprise overexploiting a fishery, but such overexploitation might carry significant costs to marginalized artisanal fishers. Since WTP and WTA are the proper measures in a CBA, the existing wealth distribution will affect the results; i.e., richer people can bear greater costs (remember that the Kaldor-Hicks criterion only states that winners could *in theory* compensate losers . . . not in practice).

There are other pros and cons to CBA that we will discuss in subsequent chapters (see Pearce 2003, Turner 2007, and Keohane and Olmstead 2007 for fair treatment of the subject). Biodiversity, sustainability, and conservation issues might be too complex to be "solved" by a technical tool such as CBA, but as Kerry Turner (2007) says, "taking decisions without any formal decision support system [something like a CBA] would not seem sensible either." Warts and all, when used correctly CBA can go a long way toward informing complex environmental decision making.

1 Coffee

2 Bieber

3 Whale oil

4 Demand and WTP

5 Park entrance fees

6 Supply curves

The Economist's Punch Line

Supply and Demand

In the last chapter we saw that Isabella could get $1,200 for the crop and timber she was going to produce on converted forestland. After subtracting the costs, she was going to make a profit of $400 over the first year. But why was the profit $400 and not some other value? The short answer is that we made it up, but in the real world the returns to farming depend on the price a farmer can charge for a ton of grain, a liter of milk, etc. And those prices have to do largely with *the market* and the market forces of supply and demand. Supply and demand are the two basic forces that determine the price for things we can purchase in the market, but how and why does this work?

Understanding supply and demand is critical to our understanding of how individuals make decisions that affect biodiversity,

The market, when explaining general economic concepts, is not typically a reference to a particular place like the local fish market or John's Corner Deli. When economists refer to *the market* they typically mean the entire universe of places where buying, selling, and trading occurs (e.g., stores, Internet sites, auctions).

such as the decision of where and how often to fish, when to convert a grassland to a forest plantation, or whether to poach a tiger for the medicinal-products market. One fascinating example of the supply-demand story related to conservation deals with whale oil in the 1800s. Oil from the sperm whale (*Physeter macrocephalus*) and the North Atlantic right whale (*Eubalaena glacialis*) was the major fuel for lighting in North America and Europe throughout the 1800s.[1] As you probably know, we whaled the heck out of these species to meet our appetite for lighting fuel and candles, to the extent that there was "an increase in consumption beyond the power of the fishery to supply."[2] Put another way, there was an increase in demand for whale oil. But as whales became more scarce and therefore more costly to hunt, fewer whales were brought to market—so the supply of whale oil was limited. With high demand and low supply, prices for whale oil greatly increased. Peak prices of about $1,900 per barrel (in 2014 USD) meant that whale oil in the late 1800s was approximately 19 times more expensive than current prices for conventional oil.

Kerosene, natural gas, and refined petroleum products eventually became cheaper alternatives to whale oil. The high

---

[1] To learn more about sperm whale oil, try reading Simon Armitage's poem "The Christening" rather than *The Whale,* also known as *Moby-Dick*, which is dreadfully long and not as exciting as this book you're reading now.

[2] Thanks to Ugo Bardi at http://www.resilience.org/ for pointing to the original resource: the 1878 Alexander Starbuck book *History of the American Whale Fishery: From Its Earliest Inception to the Year 1876.* See also Coleman 1995. Sidenote to this footnote: *Whale* as a noun is a beautiful creature, but as a verb it means to slaughter that beautiful creature. What other animal's name can you turn into a verb meaning to slaughter that animal mercilessly? Thanks to Andrew Balmford for the suggestions of *fish* and *seal.*

costs of pursuing the remaining whales, combined with diminishing demand for whale oil as a result of cheaper substitutes, lowered hunting pressure and allowed some populations to recover from historical lows, although today the North Atlantic right whale is still endangered and the sperm whale is listed as vulnerable. Current debates surrounding whaling have less to do with supply and demand and more to do with issues of cultural heritage and national sovereignty.

In addition to this fascinating historical example, the concepts of supply and demand are pertinent to a whole suite of current conservation issues. An understanding of these forces will also be critical when we start to think about estimating values for all the stuff that we cannot buy in the formal market, like biodiversity: supply and demand are fundamental to the valuation techniques covered in Chapter 5. But for now, let's start with the basics.

The Nuts and Bolts of Supply and Demand

Studies show that if you put an economist in front of a blackboard it will take about 42 seconds before she draws a graph showing supply and demand curves.[3] It's easy to poke fun at the predilection of economists to draw supply and demand curves (we make fun of Brendan for this all the time). But these graphs

---

[3] Studies don't show this, but don't you love it when authors glibly claim "studies show" without saying which study or how it was done? Anyway, we could easily crowdsource the data by setting up a website and soliciting students from all over the world to note how long it takes their professors to draw these curves in their first microeconomics class. Thesis project, anyone?

are incredibly useful for understanding many of the basics of economics, including key points that relate economics to conservation. In broad strokes, supply and demand is all about understanding how a lot of sellers (who supply) and buyers (who demand) behave in aggregate. But what does that mean?

Well, let's talk about coffee for a minute, because that is what we are drinking right now. Taylor goes for decaffeinated lattes, Robin is a straight filter-coffee man (but will take a latte when the price is right), and Brendan likes espresso in all its forms (including in a pumpkin-mocha macchiato—which is like a hot sundae—despite the grief he gets from Taylor and Robin). For those of you who don't drink coffee, these are just different manifestations of coffee-based drinks that come at different prices. In short, we demand coffee. We are currently writing at a café that offers espresso drinks for $5 per cup. At that price Taylor and Brendan are going to buy. Robin thinks that is a ridiculous price for coffee (but he doesn't have any kids waking him up at 6 a.m. either), so he goes for a basic drip coffee. So at $5 per cup, two espresso-based drinks are demanded. But what if the price falls to $2.50 per espresso-based drink? In this case, not only is Robin now in for an espresso, but Taylor buys two. If we graph this scenario (**Figure 3.1**) we can see there is a negative relationship between the cost of a cup of coffee (y-axis) and the quantity demanded (x-axis). If we can extend this idea we can think about all of the people demanding espressos such that as they get cheaper, more are demanded. This is the general shape of the demand curve. Higher demand is brought about because as the price gets lower, some buyers will buy more (Taylor) and also because additional buyers will "come into the market" at the lower price because they think the good is worth this price but not a higher one (Robin). So each person

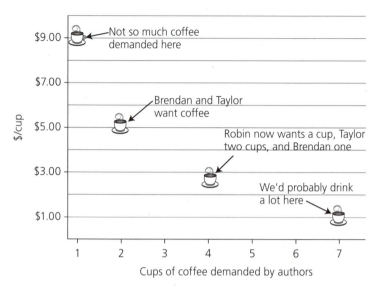

FIGURE 3.1 Demand curve imagined for coffee at a Whidbey Island, Washington, coffee shop where we wrote this section, showing the classic assumption that as price goes down, demand goes up.

has their own demand curve, their own idea of how much of a good they would buy at a given price. The market is concerned with aggregate demand—so when you take all the demand curves of all the actors in a given market (say, for coffee), you get the aggregate demand curve. This is the curve that helps to define the actual market price. But in order to get to price, we also have to understand the supply side.

The supply curve acts on the inverse rationale. The more money that buyers are willing to pay for a good, the more willing suppliers are to produce the good and the more a given supplier will produce. For example, imagine you own a coffee shop selling

cups of regular coffee at $2.50 each. If some pop star, say, Justin Bieber,[4] suddenly endorses the drinking of espresso over drip coffee, many people will be newly willing to now cough up $5 for their drinks. What will happen next? Well, for one, you might produce more cups of coffee (because now you can hire a helper to help make the extra coffee). But someone else might get into the game of selling coffee. Your friend might say, "Well, I couldn't afford to make coffee for $2.50 per cup, but now if I can charge $5 a cup, I'll start selling coffee, too." This same shift could happen even if Bieber weren't involved, because more places will start selling coffee if they see that other firms are making a profit.

Look at the simplified supply curve in **Figure 3.2**. As the price p (say, of our coffee) goes up, the more cups q will be supplied to the market. Considering the supply and demand curves together, what's going on where the supply curve and the demand curve meet (q^*, p^*)? Well, that's where both the buyers and sellers are happy. This point is the equilibrium, where all the cups made are consumed and there is no further demand for coffee at that price. In ecological terms, this is a stable state. To look at it another way, consider what happens to either side of this point. At any price above this point (higher prices), suppliers are of course more than happy to produce even more coffee, but demand is lower. (Even Taylor will give up his latte eventually). So coffee goes unsold, and coffee shops likely reduce their prices. At lower prices (below p^*) everyone starts demanding lattes, but limited supplies make for lines around the block. With lines around the block, some people will be willing to pay more for their caffeine somewhere else, and therefore the prices rise.

---

[4] As we go to press, Bieber's sway over popular culture is rapidly diminishing, so feel free to swap out The Bieb for one of your favorite pop stars, actors, sports figures, etc.

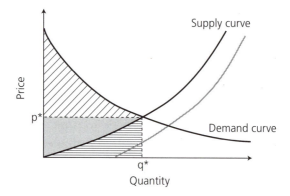

FIGURE 3.2 Simple supply and demand curves. The market clearing price is given at point (q^*,p^*), which is where supply equals demand. The top triangular area with diagonal hatching represents consumer surplus, the light-gray area represents producer surplus, and the area with hoizontal hatching represents the cost of production. The gray supply curve represents a situation similar to the global coffee glut of the late 1990s and early 2000s, when the number of suppliers and coffee on the market greatly increased in a short time period, thereby increasing supply and crashing the price of coffee (see Figure 3.3).

Economists are infamous for squeezing endless meaning out of these simple curves and the economic theory underneath them, but there's at least one other crucial lesson in them for conservation. What's the deal with the top, triangular area with diagonal hatching in Figure 3.2? This area represents what all of the consumers were willing to pay minus the cost that they actually pay for the good. In our example above, Taylor and Brendan were willing to pay $5 for a cup of coffee, but they got their cup for $2.50. So we valued that cup at $5. It would have brought us $5 worth of utility, but we only paid $2.50. Bonus! This area in the graph is called the consumer surplus, because

lots of consumers are getting their good at a lower price than they would have been willing to pay for it. What about the light gray area? This area is called the producer surplus, because lots of production would have occurred at a lower price (low part of the supply line), but all producers receive the higher amount of $2.50 per cup. Therefore, the light gray area is the surplus benefit to producers, much like the top area was an additional gain to the buyers. The gross return for producers is given by the price multiplied by the quantity sold ($q^* p^*$, or the light gray *and* horizontal hatches areas). The cost of producing q^* units is just the area with horizontal hatching. (We'll deal a bit with the general shape of these curves, how they change, and what it means to move along these curves in **Box 3.1**).

BOX 3.1 **The Shapes of Supply and Demand Curves**

So we described why the demand curve slopes downward and why the supply curve slopes upward, but what determines the actual slope of these curves? Well, a lot goes into this. Remember that the aggregate supply and demand curves are a function of all the actors in the market buying and selling, so there are a lot of factors and motivations that affect both the supply and demand sides. Price is one of them, but the number of buyers and sellers, the substitutes available, income levels, resources available for production, tastes, preferences, fads, etc.—all these things affect how goods are supplied and how many are demanded.

But let's just think about a few different options for the shape of supply and demand curves. In the figure on the next page, panel A shows your standard demand curve. It is shaped to show diminishing marginal returns. Think about pants. If you have no pants, it means you have no pants, but it also means that your first pair of pants give you a lot of benefit (p is great when $q = 1$). By the time you get to 11, a 12th pair of pants delivers only a small amount of additional benefit. You can look at it backward as well. As you start losing pants (going from right to left), the benefits of the remaining pants gets much greater. This is really typical of a lot of the goods we are concerned with in conservation. As ecosystems, clean water, clean air, and local woodlands become scarce, we typically care about the remaining ones at an increasing rate; that is, our values for them get larger. Panel B shows the standard supply curve sloping upward at an increasing rate for similar reasons the demand curve sloped downward at a decreasing rate. As the price increases, suppliers can now afford to employ more workers and to buy more fertilizer and new machines. But a producer will typically employ the most productive factors first. Think about the coffee farmers who put the most productive lands into coffee first and then over time expand to more marginal lands. Also think about the example of opportunity costs of protected areas (Box 2.1). There we learned that the cheapest lands are usually put into protection first; then it gets more expensive for each additional unit of protection when more valuable land is considered for conservation.

(continued)

BOX 3.1 continued

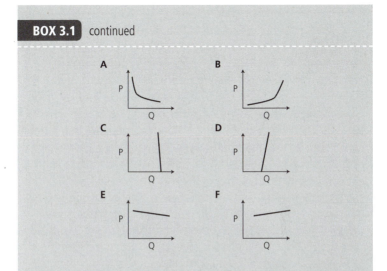

Panel C shows a demand curve that represents what economists call inelastic demand. What does that mean? Well, let's look at the curve and move backward from right to left. You can see that a great change in price does little to affect the quantity demanded. Think about oil, alcohol, rice, World Cup football tickets. All of these goods behave relatively inelastically, because the demand does not change much even if the price rises significantly. Environmental benefits begin to act inelastically when the supply level is low. Think about clean water. When supply is scarce, a large change in price doesn't change demand much, because clean water is a basic necessity. The elasticity of a good (technically, price elasticity of demand = Ed) is calculated by this equation:

$$Ed = (\% \text{ change in quantity demanded})/(\% \text{ change in price})$$

If Ed is between −1 and 0, the good is considered inelastic.

When E_d is less than –1, the good is considered elastic. Look at panel E. It shows a good for which a small change in price greatly affects its demand. Demand for a particular soda can act elastically, such that a small change in the price of soda A shows a relatively larger change in demand: $E_d < -1$. One reason for this is because there are a lot of substitutes for soda. If the price of soda A increases by 10%, the demand may drop by more than 10% because you could buy soda B, soda C, water, juice, etc.

Panels D and F are two supply curves that demonstrate the price elasticity of supply (E_s). In panel D we have relatively inelastic supply, meaning that even a large change in price only brings about a small change in supply. This situation generally reflects a good that has limited availability in its raw components in the production process or the time and knowledge of the producers. For example, some goods require rare-earth elements or a long lead time of skilled labor to produce, such that even a large change in the price only instigates a few more units produced. Panel F shows the opposite, where a small change in the price that a given good can fetch stimulates a great change in the supply of the good. Here the skills, technology, and materials are readily available. To calculate the price elasticity of supply we use

E_s = (% change in quantity supplied)/(% change in price)

If E_s is < 1 then the good is considered inelastic; if $E_s > 1$ the good is considered elastic. When E_d and E_s are equal to 1 this case is called unit elastic, such that a 1% change in price represents a 1% change in supply or demand. Right now Taylor's effort for this book is highly inelastic, such that regardless of how much we raise his royalties he still supplies very little.

Understanding producer and consumer surplus is essential for understanding how economists conceive of economic value. At the end of the last chapter we hinted at the difference between economic and financial values. Supply and demand curves now make this difference graphically clear. Financial value is what is traded in the marketplace, equal in this case to the total number of coffees sold q^* times the price for each cup p^*. But as we just saw, this is an underestimate of total value; there were a lot of folks who would buy at prices higher than the market clearing price, and a lot of folks who would produce at prices lower than the market price, such that each gains a surplus of value (or benefit, utility, whatever you want to call it). These consumer and producer "surpluses" are economic values. And remember, cost-benefit analysis is concerned with economic values.

Remember Xavi from the last chapter? He would have paid $250 for Isabella to leave her parcel as forest, despite the fact that the bushmeat he gained from hunting this parcel would only fetch a price of $200. Xavi experienced a surplus of utility from the forested parcel (based on his individual demand curve). What about Isabella? She got $p^* \times q^*$ ($700) for selling her crop, and that represented a net return of $300 for her crop (not the timber sales). That's her profit. If we knew what her supply curve actually looked like (in relation to p^*), we could get to her producer surplus. These economic values are what go into cost-benefit analysis, because they incorporate a truer sense of value than simply the financial returns.

Think about social benefits again from the last chapter; these are the benefits accruing to society at large and also very relevant to conservation issues. We can estimate a benefit curve for social values. A benefit curve is akin to a demand curve, where each

additional unit of some benefit carries less marginal utility. That sounds like a lot of jargon, but what if we say it this way: Your first dessert after a meal can be great, but how much do you actually enjoy the second dessert? Would you even enjoy a third? You see, each additional unit carries less benefit. So a benefit curve typically shows diminishing marginal utility (gain) with each additional unit of consumption. In the next few chapters we'll talk about the techniques used to consider the benefit curves for social values, but this introduction to supply and demand will be important to remember.

> The concept of "marginal" is important in economics. Put simply, the **marginal cost** of something is the cost to supply or purchase a little more of that something. The marginal benefit is the benefit a consumer gets from consuming (or enjoying) a little bit more. There are no technical guidelines for what constitutes a marginal change. But you could imagine that when a tree falls down in a 1-hectare plot of forest, that ecosystem experiences a marginal change. When someone cuts down three-quarters of that plot . . . well, that's not so marginal.

Supply, Demand, and Conservation

On one level, the impacts of supply and demand are fundamental to most conservation issues. Think about it:

Conservation is about resource decisions regarding scarce goods. For example, should land in the Great Plains region of the United States be devoted to protecting the endangered black-footed ferret (*Mustela nigripes*) or to raising cattle?

Resource decisions are sensitive to price. For example, the price a farmer gets for selling a cow determines how many cows to produce and therefore how much land to ranch.

Prices are driven by supply and demand in the marketplace. For example, how much are people willing to pay for beef as opposed to other things they like to eat?

Changes in supply and demand affect the benefits and costs to people and society. For example, increased demand for beef might mean more ranch lands and less future habitat for the black-footed ferret (*Mustela nigripes*), less land for bird watching, and so on.

Voilà! That's why we couldn't avoid putting this topic in the book. Supply and demand for various commodities and marketed goods reflect the relative value that people place on those things. But perhaps more importantly, supply and demand dynamics also operate outside of traditional markets. Sometimes we need to think about the values of non-market goods or things we typically don't buy, trade, or sell in markets; for example, values associated with securing a future for the black-footed ferret, experiencing wild nature, or the importance of forests in improving downstream water quality. Indeed, most of the non-market valuation techniques that we will investigate in Chapter 5 operate by reconstructing these curves in ways not clearly accessible from existing market information. And so understanding the demand for open spaces, national parks, clean air, polar bears, and so on helps us to make decisions and allocate resources toward supplying these social benefits.

Checking back in on Arden, let's think about what happens to Isabella's farming decisions under shifting supply and demand. We'll now assume that the crop she produces is coffee. In the last chapter, she expected $700 per year in revenue from her new coffee farm. But in the early 2000s, increasing demand

for coffee fueled an expansion of production in Brazil and Southeast Asia; this increase in supply overshot the demand, so globally we had too much coffee. How would such an excess of supply be reflected in our familiar graphs of supply and demand? Because there is so much coffee on the market, the prevailing conditions of the previous supply curve have changed. Suppliers are now forced to revise downward the price at which they would be willing to produce various amounts of coffee, since they know that their competitors are doing the same in light of the excess coffee "flooding" the market.

All of this translates into a shift to the right of the supply curve; producers now must be willing to sell at a lower price for any given quantity of coffee (see the light gray supply curve in Figure 3.2). When that happens, look at how the equilibrium price (or in economic jargon, market-clearing price) is now lower than the previous market price (in Figure 3.2, where the light gray supply curve intersects with the demand curve). And sure enough, if we examine real coffee prices this is what actually happened: the price of coffee fell almost 60% between 1998 and 2001 (**Figure 3.3** on p. 54). With a similar drop in price, Isabella's expected revenue from coffee declines from $700 per year to $300 (**Figure 3.4** on p. 55). With all else the same, her profit from converting her forest would also fall from $400 to exactly zero (handy, huh?). The opportunity cost of the conservation of forest on Isabella's land has now disappeared— along with her financial incentive to convert. Based only on this cost-benefit analysis, we can imagine that Isabella is unlikely to convert her forest. And so we see the strong link between how the price of a globally traded commodity, coffee, can affect very local decisions by landowners. One example of this in the real world is the evidence that deforestation in the Brazilian

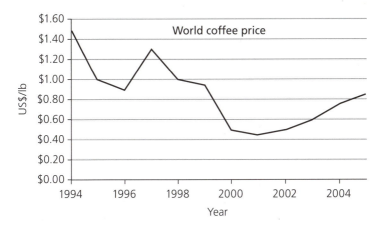

FIGURE 3.3 Response of global coffee prices to the oversupply in the late 1990s and early 2000s. (Source: International Coffee Organization.)

Amazon has in the past been largely driven by the market prices of soy and timber (remember, from Chapter 1?).

Demand curves alone, even without an accompanying supply curve, can help us understand conservation problems and point us to solutions. As an example, take wetlands, which are estimated to have been reduced by 50% from their global extent in 1900 (Moser, Prentice, and Frazier 1996). Wetlands are the feeding, breeding, and nursery grounds for countless numbers of fish, amphibians, reptiles, and mammals. Thousands of bird species rely on wetlands as the proverbial stepping stones on their long-distance migrations, including the globally threatened (but highly promiscuous) aquatic warbler (*Acrocephalus paludicola*), which relies entirely on wetlands as stopover grounds on its 4,000 mile journey from western Africa to breeding grounds in eastern Europe.

Current landscape   **Convert?**

☐ Isabella   ☐ Xavi   ▨ Forest   ▦ Agriculture

BENEFITS	COSTS		
Private benefits	*Private costs*		
Timber revenue........$500	Timber harvest........$400		
Agriculture revenue..$300	Crop planting,		
	harvest...................$400		
Social benefits	*Social costs*		
	Loss of hunting and		
	aesthetics for Xavi..$250		
		Difference	Ratio
TOTAL......................$800	**TOTAL.....................$1,050**	($250)	0.8
Private total.............$800	Private total.............$800	$0	1.0
Social total..................$0	Social total...............$250	($250)	0.0

FIGURE 3.4 Arden's landscape, showing Isabella's decision-making criteria for whether to convert forest parcels to coffee. This time, global declines in coffee prices have revised her cost-benefit analysis and made the decision less clear.

- -

Apart from their biodiversity values, wetlands are also vital to the supply of ecosystem services such as water regulation and waste assimilation; we'll provide a more complete list of ecosystem functions and services in the next chapter. The conservation community has been so concerned with wetland

losses that in 1971 the Ramsar Convention was signed by 160 countries to foster national and international collaboration for their protection. But do we really value them? And how does this value change with the quantity supplied?

To answer this question, Luke Brander and colleagues (2006) reviewed all of the studies they could find that investigated people's willingness to pay for wetland conservation. Looking at nearly 200 studies that had elicited people's values for benefits like flood control, species habitat, biodiversity, and sediment retention, they found that people's WTP for wetlands across the globe declined with the size of the wetland they were asked about. So it looks like people display a classic demand curve for an environmental good such as wetlands, just as they do for coffee. As the quantity of the good (here the size of a wetland) increases, the marginal benefit (measured by WTP) of the good decreases. (Take a look at Figure 3.2 to remind yourself.)

Understanding supply curves can be just as helpful to conservation. As we have seen, the supply curve is essentially the cost of supplying any given good. In conservation we are constantly working to supply more things—more forests, more species, more free-flowing rivers—under real-world resource constraints. It might seem strange to think about the cost of supplying species, but in most cases it costs money, time, and effort to protect species and ecosystems. And if you spend that effort on some species, there will be less available to protect others. So cost matters.

David Wilcove and Linus Chen (1998) looked into this issue for endangered species in the United States. They looked at the costs of alleviating threats to different endangered species and found huge differences between, say, the cost of monitoring the harm to native freshwater mussels caused by the invasive Asiatic

clam and the cost of restoring natural fires to landscapes inhabited by the fire-dependent and endangered red-cockaded woodpecker (*Picoides borealis*). In other words, different species have different supply costs. If we take these data and arrange them in increasing order of cost to alleviate threats, we get a supply curve like that in **Figure 3.5**. These data make clear that it is much more expensive, on a per-species basis, to alleviate the threats to birds, reptiles, and amphibians than it is to alleviate the threats to other species groups. Imagine you were a reserve manager and your job was to alleviate the threats to the largest number of endangered

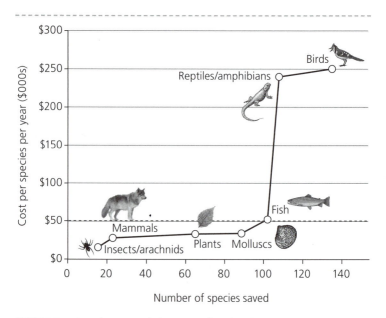

FIGURE 3.5 Supply curve of the costs of undertaking conservation actions to alleviate the threats to endangered species in the United States. (Derived from Wilcove and Chen 1998.)

species with a fixed budget of, say, $5 million per year. You could most efficiently achieve this goal by understanding the marginal supply curve shown in the figure. In this case, with $5 million you could "supply" 100 species (the straight dashed line in Figure 3.5), although this means not taking direct action on the protection of reptiles and birds.

Let's consider one more supply-side example by getting back to coffee. In some parts of Indonesia, coffee plantations lead to a lot of soil erosion, fouling up waterways and dams and causing ecological and economic damage downstream. To counter this, the decision makers in Indonesia considered giving farmers incentives to practice conservation agriculture, carefully planting and managing coffee in ways that reduce soil erosion. These techniques carry costs—less environmentally harmful production practices might actually produce less coffee per unit area, and they also entail additional management costs. So it is unlikely that many farmers would adopt these practices without some financial incentives. But just how much would you have to pay them? Kelsey Jack and colleagues (2009) asked this very question: How much would it cost to pay farmers to undertake conservation agricultural practices and hence reduce on-farm erosion and downstream ecological and economic costs? Put in more economic terms, their question was, "What is the cost of supplying a unit of erosion control?"

Just as in the endangered species example, Jack was essentially after a supply curve. Since different farmers have different opportunity costs (remember those from Chapter 2?), each is probably willing to enroll in the program at different payment levels. In order to "reveal" these differing opportunity costs the researchers held an auction, which basically asked the question, "How much would we have to pay you to undertake conservation

agriculture?" Sure enough, farmers had widely different answers. Arranging their answers into a supply curve, the research team could demonstrate the costs of supplying varying amounts of erosion control. Take a look at **Figure 3.6**. Farmers on about half of the available hectares (up to 40, more or less) would accept a relatively low payment to enroll. After that, though, things get expensive; some farmers see high opportunity costs of enrolling, so they are willing to do so only for high payments. With that supply curve the World Agroforestry Centre (ICRAF) was able to "buy" erosion control most cost-effectively by buying the cheapest parcels first. Cool, huh? Here an exercise to elicit the opportunity

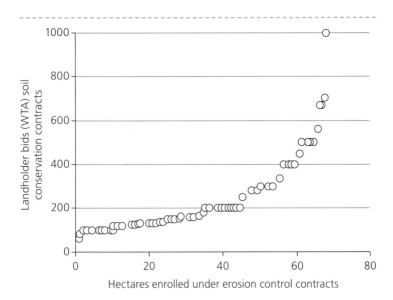

FIGURE 3.6 Supply curve of the cost of reducing erosion through improved agricultural techniques in Indonesia (Redrawn from Jack et al. 2009).

costs of many farmers revealed a supply curve that helped to cost-effectively improve an environmental outcome.

OK, so we've seen how supply and demand curves can be estimated and used individually for things we care about in conservation. Here's an example of bringing both sides together, done by Robin Naidoo (yes, one of the authors of this book actually did some decent research, but this was some time ago) and Vic Adamowicz (2005). What's cool about this study is that they estimated both the supply and demand curves to inform how much a tropical forest reserve should charge visitors to conserve the greatest number of bird species. First they looked at the marginal benefits of increasing numbers of bird species (for tourists to view) at Mabira Forest Reserve in Uganda. They wanted to answer the question "How much is an additional species sighting worth to tourists?" This is equivalent to asking what is the *demand* for an additional sighting; i.e., the demand curve. They estimated this demand curve by carrying out a choice experiment—basically a fancy way to ask tourists how much they would be willing to pay to see more birds. We'll learn about choice experiments in Chapter 5.

They also looked at costs of supplying this bird diversity; i.e., the *supply* curve. The supply curve was derived by calculating the opportunity cost of forest conservation for each hectare of forest—estimated by multiplying the likelihood that any parcel of forest in the reserve would be converted to agriculture by the expected net return to agriculture in the region—and then converting hectares of forest conserved into bird species conserved via a species-area curve. By comparing the cost and benefit curves, they found that the benefits generated by current entrance fees would be enough to offset the opportunity cost of conserving 80% of the forest's bird species. This would rise

to 90% if entrance fees were set at levels that would maximize revenue (as with many protected areas in developing countries, entrance fees are currently set too low). This research has provided us with a clear example of how understanding supply and demand curves can practically inform on-the-ground conservation decisions that affect biodiversity and the tangible benefits it provides to people.

Can We Conclude That Supply and Demand Are Important for Conservation?

Essential to the future of conservation is our understanding of the effects of land-use change on biodiversity, ocean management on marine ecosystems, population dynamics under changing climates, the spread of invasive or exotic species, threats to protected areas . . . and the list goes on. Here we are suggesting that it is also critical for conservationists to have a basic understanding of how the forces of supply and demand can affect conservation issues. In our economics-speak, in this chapter we saw how opportunity costs and willingness-to-pay helped us to construct supply and demand curves, and how these curves helped us to understand economic values. In non-economics-speak, in this chapter we've seen how they help us with on-the-ground interventions such as charging park entrance fees, optimizing expenditures on species conservation, and mitigating erosion, and they also help us to understand higher-level issues such as society's demand for conservation, wetlands, ecotourism, and so forth. These benefits from nature are often referred to as ecosystem services, and they are the subject of a detailed examination in the next chapter.

1 Ecosystem services

2 Public goods

3 Private goods

4 Carbon storage

5 Ice Cube

Ecosystem Services

Where Nature Meets Human Welfare

The rapper and actor Ice Cube once said, "Life is nothing but ladies and money." Actually, his words were a little more colorful than that, but anyway, the point is: he's wrong. He completely forgot about the benefits we derive from ecosystems.[1] If you've ever had a glass of water in New York City (or many other cities), taken a hike to a waterfall, eaten tomatoes, drunk coffee, or bought fresh fish, you've enjoyed the benefits provided by nature. That's because ecosystems and the species in them purified that water, provided the nice calming scenery, pollinated those tomato and coffee plants, and produced those wild fish. These are nice luxuries, but some people are utterly dependent on ecosystems for their livelihoods—think of people in fishing communities in developing countries or those who harvest wild foods from the forest.

The wonky term for benefits derived from nature is *ecosystem services,* and it is a popular concept in conservation these

---

[1] Of course he forgot a lot of things, not just the benefits we get from nature but also love, security, friendship, soccer . . . and probably a few others.

days[2] because it highlights the ways in which conserved nature is valuable to our lives. We traditionally don't account for these benefits in decisions we make, so we risk losing opportunities to improve our lives and our economies by managing our ecosystems to continue providing services. That is an argument for conservation-based economics and human livelihoods: Ecosystem services offer conservationists new ways to motivate conservation (it's in our own interest) and new ways to finance it (by having those who enjoy the services pay those who provide them).

This sexy and timely idea is, at its foundation, an economic one. (When's the last time *that* happened?) It's about recognizing how functioning ecosystems provide for our needs, about the economic and even monetary value of such resources, and about making decisions under scarcity (this last one is the very definition of economics). So we need to understand the economic basis for ecosystem services—that's what we tackle in this chapter. We first jump to some fundamental stuff on the nature of goods themselves, and then we get to ecosystem services and their economic considerations.

The Nature of Goods

Perhaps you've heard of the tragedy of the commons? You know, where lots of folks are grazing their livestock on the same pasture, and they race to overgraze it, thus exhausting the resource and hurting themselves in the long run (Hardin 1968)?

---

[2] If you want to read the seminal work in this field, pick up *Nature's Services* edited by Gretchen Daily (1997).

That's the famous example of common-pool or open-access resources,[3] which are a pervasive concern in conservation. But for economists this concept is way too simple and intuitive, so they have extended this idea into four categories of *goods*, based on two properties: whether that good is "rival" and whether it is "excludable."

> A **good** is something that provides a benefit, or utility, to a person or group of people. Goods can be tangible, like a mango, or intangible, like information.

Let's start with pizza. If Robin eats a slice of pizza, that pizza is no longer available for Brendan to eat. That makes it "rival" (if Robin consumes it, Brendan can't). Also, if Taylor buys that pizza slice, he can prevent both Brendan and Robin from eating it. That makes it "excludable" (Taylor can prevent his coauthors from accessing it). These kinds of goods, both rival and excludable, are called private goods.

Now let's think about a contrasting case: a lighthouse, which provides safety at sea and protection from the crashing of ships, cargo, and passengers against the coast. Is it rival? No, because one person's use of a lighthouse does not mean there is less for another to use. Is it excludable? No, because it is nearly impossible to keep someone from using a lighthouse. Lighthouses, and other nonrival, nonexcludable goods, are called public goods.

The concept of goods can be a bit of a mind-bender, we know. Check out the handy table in **Figure 4.1** on p. 66 to see all combinations of rival and excludable goods. And let's fill in the other two boxes.

---

[3] There actually is a distinction here between common-pool and open-access resources, but for this discussion you can consider that open-access resources are common-pool resources with absolutely no management or governance.

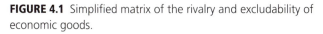

	Excludable ("I can prevent you from accessing it")	Nonexcludable
Rival ("If I use it, there is less for you")	**Private good** (crabs or fish)	**Common-pool resource** (public grazing lands)
Nonrival	**Toll or club good** (copyrighted information)	**Public good** (stable climate)

FIGURE 4.1 Simplified matrix of the rivalry and excludability of economic goods.

First, a toll or club good is one that is excludable but nonrival; an example is information. My use of information—say, learning a new methodology for surveying large cats in the wild—does not leave less of that knowledge for you. In fact, using that information might actually improve and expand it by refining the methodology further. But if that methodology is written up in a journal with large subscription costs, such that you can be excluded from using the information, then that is a toll good.

The final box of goods captures the tragedy of the commons idea we started with. These are goods that are rival but nonexcludable. One user can't prevent others from using it, but each use leaves less for others. You can see how these common-pool/open-access resources are prone to tragedy.

Ecosystems provide us with goods and services that fit in each of the boxes in Figure 4.1. Quiz time! See if you can place these: timber cut from forests, fish harvested from the open ocean, medicines harvested from rainforests, skin protection

from the ozone layer, stable climate regime, wildlife viewing in a national park.[4]

If you actually did take this quiz, you probably realized that this four-box breakdown of goods is probably too simple. In reality there's more of a gradient of rivalness and excludability. For example, think about the Dungeness crab we're eating while we write this. If Robin catches a crab, steams it right on the beach, and eats it,[5] he leaves less crab for Taylor (rival), and he keeps Taylor from eating it (excludable). That makes the crab a private good. But the crab fishery itself is more like a common-pool resource—rival but not excludable—and therefore prone to overfishing without good management.

Regardless of these occasionally fuzzy boundaries, the matrix in Figure 4.1 is a useful framework for thinking about the benefits we derive from nature. The following in particular are a couple of the most important lessons for conservationists.

First, whether a good is rival or nonrival often depends on how heavily it's being used. In coastal Mozambique, for example, local people fished the seas at low levels for centuries without really degrading the resource. An offshore upwelling

---

[4] Here's what we think: Timber cut from forests, private good; fish harvested from the open ocean, private good (fish) or open-access resource (ocean); medicines harvested from rainforests, private good (if made and sold in markets) or public good (if medicine is information and freely available) or toll good (if the information of medicinal benefit is patented and access to it is restricted); skin protection from the ozone layer, public good; stable climate regime, public good, but becoming congested; wildlife viewing in a national park, public good (if free) or toll good (if charged).

[5] Due to both the relish with which this event took place and the sheer number of crabs eaten, this event has become known as "The Gluttony at Holmes Harbor."

provides rich sources of nutrients for both shrimp and tuna runs. Like Robin's crab, the fish is strictly a rival good (if you catch it, I can't). But for decades there was enough for all, so in practice, fish behaved more like a nonrival good. With the arrival of commercial fishing fleets, however, harvest rates have risen and fish are becoming rival (i.e., my fishing degrades the resource for you).

The second important lesson is that private goods are easier to sell in the market; that is why they are also called market goods. Remember that markets are places of exchange, and these characteristics (rival and excludable) provide clear incentives for people to produce these kinds of goods and exchange them in markets. These days we think mainly in terms of monetary transactions—that is, buying things for money—but think about barter exchanges for a minute. If I come to the market trying to trade you sunlight (nonrival, nonexcludable) for a new goat, I'm not likely to return home with a new goat (or even an old goat, for that matter). When trading, I need to give you something I can keep from you (excludable), otherwise you would just use mine. So when I offer you a new shovel, you are happy to turn over that goat.

Here's the point. Many ecosystem services are not private goods. They aren't easy to "own" or to sell on the market. The result is that, in general, we tend to under-invest in supplying them, because it's harder to capture a profit by doing so. In fact, the Millennium Ecosystem Assessment (2005)—a huge effort by 1,000 scientists over 5 years to assess ecosystem services—found that 60% of them are declining. The 40% that weren't declining were all private goods. That's telling you something.

Keeping this framework in mind will help us adopt an economist's mind-set when we consider all the stuff that nature

does for humans—and the struggle to manage them in the absence of obvious incentives to do so.

What Are Ecosystem Services?

In rough terms, ecosystem services are the things that ecosystems do that benefit us. To quote Daily et al. (2000):

If properly managed, [ecosystems] yield a flow of vital services, including the production of goods (such as seafood and timber), life support processes (such as pollination and water purification), and life-fulfilling conditions (such as beauty and serenity). Moreover, ecosystems have value in terms of the conservation of options (such as genetic diversity for future use).

But wait, you say, we've been talking about managed ecosystems from the beginning, with things like Xavi's hunting or tourists enjoying the birds of Uganda. Yep, that's right. But *ecosystem services* is the umbrella term for all the things ecosystems do that benefit people. In fact, there is a debate among ecologists and economists about the proper definition and categories of ecosystem services—but we don't need to wade into that here.[6] Instead, we want to stress three things that are really important for the purposes of this field guide.

First, ecosystem services is an anthropocentric concept—it is concerned with benefits to humans. If nobody benefits, there's no ecosystem service. Coral reefs produce fish, but if

---

[6] See Boyd and Banzhaf 2007; Fisher, Turner, and Morling 2009; and Bateman et al. 2011.

nobody is there to eat them, they are not providing a food benefit to anyone. Wetlands can purify water, but if nobody lives downstream, there is nobody to enjoy this ecosystem service. Reefs and wetlands are of course important for lots of other reasons, such as their support of many species of plants and animals, and for simply their own right to exist, but ecosystem services require people who benefit.

Second, ecosystem services are not the ecosystem that provided them. They are what the ecosystem *does.* Reefs aren't the service; fish production is. Wetlands aren't the service; water purification is. Clean water isn't the service; the process of wetlands helping to clean it is. These services produce benefits that people enjoy, like food, water, and recreational opportunities. See **Box 4.1** on p. 72 for a picture of how all these terms relate to each other.

Third, ecosystem services can be thought about in terms of their position among the goods described in Figure 4.1, which helps us explain why the market is pretty good at supplying us with private goods like timber, lobsters, and statistical distribution pillows.[7] At the same time, services like climate regulation provide benefits (e.g., stable weather patterns) as a public good: I can't keep you from using it, and my use of it doesn't affect your use of it.

Whoa! Enough with the concepts. Let's look a bit more concretely at this last example: storage of carbon by ecosystems.

In broad strokes, the carbon cycle involves plants fixing atmospheric carbon via photosynthesis, then releasing that carbon as they decay or as animals eat them and respire, which is stored again in the soil, water, or atmosphere. Roughly 40% of all carbon stored on land is in forests, and 34% is in grasslands.

---

[7] See http://www.etsy.com/shop/NausicaaDistribution and buy these pillows as birthday presents for your math-nerd friends.

In your typical forest, most of the carbon is stored aboveground in tree trunks and branches. In a grassland system, most carbon is stored underground in roots and soil. So if we cut down forests, disturb grassland soils, or burn either place, we release these stores of carbon into the atmosphere as CO_2 (mostly), sending it through the cycle again.

OK, so that's a bit on the biophysical side of *carbon storage*, but what benefits does it confer to people? Where in this is the ecosystem service? Well, you'd have to have been buried deep in a grassland's soil not to know that increasing CO^2 in the atmosphere is one of the main reasons for climate change. It and other gases trap heat and warm our planet. So storing carbon is one of the services forests and grasslands provide. A stable climate benefits us in many ways, mostly by preventing unpleasant things like increased storms, flooding by rising sea levels, shifting crop zones, and many others. A whole lot of people are busy estimating how much these problems would cost us. The average estimate of the "social" cost of a ton of carbon released into the atmosphere is about $40.[8]

> **Carbon storage** refers to the ability of ecosystems like forests to trap carbon in trees, roots, and soil. Life is largely built from carbon, so all species and especially plants contain a lot of it. **Carbon sequestration** refers to a change in the amount of stored carbon. If an ecosystem is storing more carbon over time (say, because trees are growing), it is sequestering carbon.

[8] This is roughly the median value from a guy named Richard Tol (2005), who reviewed everyone's estimates of the social damages of emitted carbon: the economic hardships of a changing climate, the costs of cleaning up after more frequent weather disasters, etc. The estimates vary a shocking amount, from a few dollars to more than $1,000 per ton. Tol's done a great job of estimating what the average values might be, given this range.

BOX 4.1 Untangling Terms for Ecosystem Services

You hear about ecosystem services everywhere these days, but often with a mixed-up muddle of terms. In particular, words are used interchangeably that aren't in fact synonyms. Here is our attempt to sort it out for you, but remember that even real live economists and ecologists can disagree on what means what.

Check out the graphic that follows. An *ecosystem* provides an *ecosystem service,* which confers a *benefit* to people, which we measure by estimating its *value.*

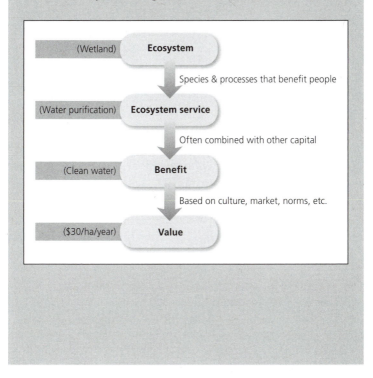

We define these terms elsewhere, but here is a summary:

- **Ecosystems** are combinations of interacting species and their environment. They do many things, not all of which are ecosystem services.
- **Ecosystem services** are the subset of these natural processes that actually generate benefits to people. Sometimes this is the production of a good.
- **Benefits** are the actual things that increase human welfare.
- **Value** is the importance of a given benefit to a person or group, measured in dollars or other metrics.

OK, an example. A wetland is an ecosystem. Among many things a wetland and its myriad species do, they purify water as it flows through. So the ecosystem service is water purification (what the ecosystem actually *does*). The benefit is clean water (the result that actually benefits people). And the value is $30 per hectare of wetland per year, which we might calculate by figuring out how much it would have cost to clean the water ourselves. Again, if nobody lives downstream, the value is $0. Clean water could also be valued using a number of different metrics. For example, "sick days avoided" might be the most meaningful metric in a situation where having access to clean water means better health outcomes for a community.

Use the figure and try it yourself to analyze carbon storage, bird watching, flood control, coral reefs, or bushmeat. (*Hint:* This list includes ecosystems, ecosystem services, and benefits, so for each make sure you start in the right place.)

If you want to delve further into this definitional stuff we'd first recommend reconsidering, but then we'd recommend reading Daily (1997); Fisher, Turner, and Morling (2009); Boyd and Banzhaf (2007); or Bateman et al. (2011).

Carbon makes for a good example, but there are hundreds of such processes that make human life possible. Some things are accomplished by a few species, like a couple of species of bats that eat crop pests. Others require a symphony of species and their environment, like diverse coral reefs that delight divers and tourists. And a given ecosystem can provide a lot of different services. For example, wetlands can provide us with goods such as food, fiber, genetic resources, potable water, energy, and ecological knowledge, as well as services such as coastal protection, erosion protection, flood protection, climate stabilization (via carbon storage), and waste assimilation.

What's cool is that economics provides us with the tools to understand the importance of these benefits to people, and the value of providing or losing them. So let's return to Arden and see how our example fits in with what we've been learning so far.

Supply, Demand, and Externalities in Arden

Remember when Isabella decided to convert her forest into cropland in Chapter 2? The first thing she did was cut and sell the timber to make a profit of $100. There were lots of other benefits provided by that forest. We've already talked about Xavi's hunting; the supply of bushmeat is another ecosystem service. But let's continue our focus on carbon storage by forests.

Linking back to our goods framework in Figure 4.1, timber in this case is a private good, and Isabella can sell it for private benefit. But carbon storage is a public good, as we've explained. Landowners like Isabella supply it, but others (pretty much everyone on Earth) benefit from the climate stability it helps to provide. That means at a given quantity of forest supplied, the

benefit of the marketed good (timber) is likely to underestimate the net value of the forest. The take-home lesson is that if we only include the marketed goods in our analysis, then society will certainly under-provide the natural systems we rely on. A huge social benefit will go un-provided, and demand for natural systems and the benefits they provide will go un-met.

This situation is known as an *externality*, another wonky term but one of the most pervasively important economic concepts for conservation. An externality, in general terms, is a cost or

> An **externality** is a cost or benefit that isn't explicitly captured in a decision or action. It is external to the decision for the person making it.

benefit that isn't explicitly captured in a decision or action. The classic externalities are typically negative things, like pollution. Pollution by a factory might affect water or air quality downstream, house prices in the area, or human health downwind. That means other people—not the plant's owners—are bearing the costs of the plant's pollution. And that is why pollution is considered *external* to the owners' decisions. Deforestation also carries negative externalities—siltation of rivers, loss of biodiversity, climate change, and so on. Isabella gets the timber revenue, but Xavi and others bear the costs of lost hunting.

Here's the point: Many ecosystem services are externalities. In contrast to pollution, though, they are positive externalities—the benefits from conserving ecosystems often accrue to others, not the owner or manager of those ecosystems. If that's true, then you might wonder why a landowner would provide ecosystem services if there's no private gain.

So how do you fix externalities? You internalize them, of course. That's really what economists call it. What it means is: Give Isabella a private incentive to provide a public good. For

example, a government might pay her to replant forest along rivers, as a way of improving the quality of drinking water for a city downstream. Arthur Pigou (1877–1959), another economist with a cool name, said that in order to fix externalities governments need to tax actions that create negative externalities and subsidize actions that create positive externalities. So, tax folks who pollute streams, and pay people whose lands purify them.[9]

What's the right price for improved water quality? That's the topic of Chapter 5. What's the best way to pay Isabella? That's in Chapter 6. Patience, Grasshoppah.[10] Let's look at and refine our cost-benefit analysis first, which will show us *why* having such payments could really change the course of land-use decisions.

Cost-Benefit Analysis and Ecosystem Services

OK, so let's imagine that a system of carbon payments has sprung to life, and let's revisit Isabella's cost-benefit analysis (**Figure 4.2**). She can now receive a payment for storing carbon of $5 per ton.[11] Looking more closely at the two parcels she's been thinking about converting for two chapters now, we see

---

[9] Obviously, there are other ways to fix externalities (like just talking to your neighbor), and sometimes using the market to fix them may backfire. We'll get to talk a bit about how markets and institutions sometimes crowd out good behavior in Chapter 8.

[10] It saddens us that many readers will need this footnote, but here it is: en.wikipedia.org/wiki/Kung_Fu_(TV_series)

[11] Five dollars is a lot less than the social cost of carbon (Tol 2005), but it is closer to the actual price on current carbon markets (www.ecosystemmarketplace.com/marketwatch/). This might be the clearest evidence of all that society undervalues public goods.

☐ Isabella	☐ Xavi	▨ Forest
▦ Sparse forest	▧ Agriculture	

BENEFITS	**COSTS**		
Private benefits	*Private costs*		
Timber revenue........$250	Timber harvest.........$200		
Agriculture revenue..$250	Crop planting,		
	harvest....................$200		
Social benefits	Lost carbon payment...$200		
	Social costs		
	Loss of hunting and		
	aesthestics for Xavi...$125		
	Social costs of		
	carbon...................$1,600		
		Difference	Ratio
TOTAL......................$500	**TOTAL....................$2,325**	($1,825)	0.2
Private total.............$500	Private total.............$600	($100)	0.8
Social total..................$0	Social total...........$1,725	($1,725)	0.0

FIGURE 4.2 Arden again, now with an operating carbon market. Isabella is considering converting to coffee only the eastern parcel, which holds denser forest. New private costs (from forgone carbon payments) and social costs (from the social damage due to a ton of emitted carbon) now enter the cost-benefit analysis.

that they actually differ in their forest cover. One has dense forest, and let's assume it holds 50 tons of carbon, while the other (perhaps on poorer soils) has sparse forest, holding only 20 tons. Coffee plantations in Arden hold 10 tons of carbon per parcel, so conversion of dense or sparse forest will result in a *net* release of 40 or 10 tons of carbon, respectively. This net change is what will affect the climate, so it is what the incentive should relate to. Right? Stare at the wall and convince yourself of this before moving on.

When there was no carbon market to consider in Chapter 2, Isabella's internal cost-benefit analysis led her to convert both parcels (Figure 2.3). But now let's add the benefits of conservation: the carbon payments Isabella will receive. In Figure 4.2, Isabella is thinking about cutting *only* the dense forest parcel. The carbon payments appear in her analysis as a cost of cutting the forest, so it goes under the cost side (40 tons of lost carbon at $5 per ton). Her agricultural revenue drops to $250 for this one parcel (reflecting a moderate price for coffee), and all other values are half of those listed in Chapter 2. We can see that Isabella's private cost-benefit analysis now indicates that it doesn't make sense for her to convert the parcel—she'd lose $100 in doing so.

Just as important, the full CBA (including the social benefits) is also negative—highly negative, in fact—because society is bearing the social cost of emitted carbon, which as we said is roughly $40 per ton. So now everyone appears better off if that parcel stays forested. Isabella's private incentives are aligned with benefits to society at large. The externality has been internalized and has changed Isabella's behavior. Nice story, huh?

OK, so what about her other forest parcel, the sparse one? Do this calculation yourself; update the table in Figure 4.2 and

see what the CBA indicates to Isabella. Go ahead and scribble in the book. Either way, it appears that carbon payments in Arden would at least slow deforestation by half, and maybe stop it entirely, depending on your results.

So, have these cost-benefit analyses been done for real ecosystems? In 2002, Andrew Balmford and colleagues reviewed all the papers they could find to try to answer the question, What returns greater benefits to society: conversion or conservation? After reviewing over 300 case studies, they found only 5 studies that estimated marginal values, included at least one market good, and had a consistent valuation technique. We reproduce a couple of the results in **Figure 4.3** on p. 80. In each case, converting the ecosystem to maximize private benefits provided lower total benefits, just as in our Arden example.

But of course it is really more complicated than these simple "net benefit" calculations. Let's think about some spatial aspects first. As we saw in Chapter 2, Isabella's decision to conserve her forest benefits her immediate neighbors in terms of aesthetics and bushmeat, and now we see that it benefits people all over the world in terms of stored carbon. The point is that different externalities affect different groups of people, in different places. Who and where depends on both the nature of the service (i.e., global, like carbon storage; or local, like bushmeat) and simply where people are. It turns out there are probably four general ways in which the benefits flow from where they are produced to where they are enjoyed, which we think is so helpful that we devote **Box 4.2** on p. 81 to this point. So, summing up the full private and social costs and benefits of an action involves tracing these effects to the appropriate places and people. Like we said, it's pretty complicated.

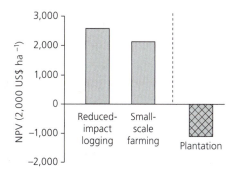

Tropical forest, Cameroon

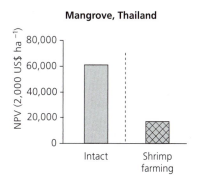

Mangrove, Thailand

FIGURE 4.3 Examples of cost-benefit analyses for tropical forests and mangroves. Units on the vertical axes are net present value—or the value in today's dollars of benefits that extend into the future. In Thailand, converting a mangrove system to shrimp aquaculture delivers only about 25% of the benefits of keeping it intact to provide sustainable timber, juvenile fish, and storm protection. In Cameroon, converting forest to a rubber plantation actually yielded a negative benefit to society. This is because the private gains from conversion were much smaller than the social costs. (Adapted with permission from Balmford et al. 2002.)

BOX 4.2 Spatial Dynamics of Ecosystem Services

There are many ways that ecosystem services "flow" from where they are produced to the people that actually benefit from them. Here are some cartoon examples of these flows. We ingeniously use the letter *P* for where the service is produced and the letter *B* for where the benefit is felt. In the first box both the provision and benefit occur at the same location (e.g., soil formation, provision of raw materials like firewood). In the second box the service benefits the surrounding landscape in all directions equally (e.g., think about native bees flying out from their native forest to pollinate surrounding crops). Boxes 3 and 4 demonstrate services that have specific directional flows. In box 3, people downslope benefit from services provided upslope (water regulation from uphill forests, anyone?). In box 4, the service flows in only one direction, as in coastal wetlands providing storm and flood protection to a coastline. (This figure is adapted from Fisher, Turner, and Morling 2009, but to be honest it's based on ideas stolen from the other two coauthors.)

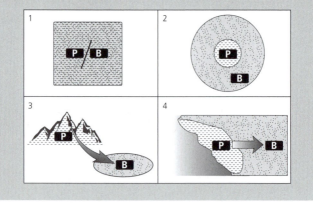

A few further considerations: What if Isabella's forest is heterogeneous in ways other than carbon storage? For example, what if the quality of bee habitat also changes considerably across the parcel? These considerations could affect her decision. Perhaps bees provide pollination services such that by keeping certain bits of habitat, they would actually enhance the productivity of Isabella's coffee crop. Perhaps conservation organizations are interested in protecting the key breeding habitats and are willing to help Isabella conserve just those areas.

These nuances make the example a bit more real and beg several questions: How can we actually enumerate the value of all these things we don't have markets for? How do we know how much to pay for carbon storage? How much additional value does having a few bees on the land give the coffee crop? And, in general, how can the interests of society mesh with the private interests of Isabella?

These questions are really about *value*. When we have a better understanding of the different values of the ecosystem goods and services (and who values them) we can make more informed decisions about the world we live in. This is what the next chapter is all about—valuing ecosystem services and the benefits that we derive from them.

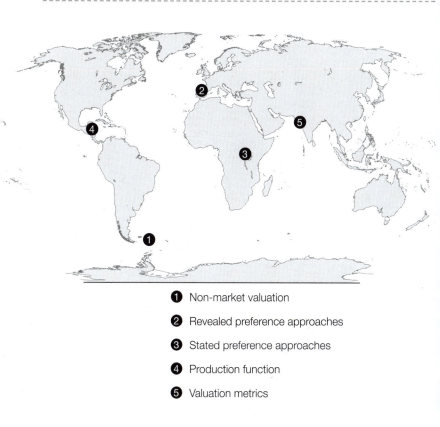

1 Non-market valuation

2 Revealed preference approaches

3 Stated preference approaches

4 Production function

5 Valuation metrics

Valuation Techniques

How to Estimate Nature's Benefits

All right, so nature provides benefits to people—so-called ecosystem services. And these benefits are one of many reasons to conserve natural ecosystems and their component species. But what are these benefits worth? How important are they compared to other things we care about in our lives? Economists tend to answer this question by valuing stuff: estimating its economic value, most often in dollar terms, and figuring out what factors lead some people to value nature more than others do. Like trying to remove a porcupine from a sock drawer, this gets tricky. It is also contentious both on scientific and ethical grounds. The alternative is akin to assuming that the value of nature's benefits is zero. So let's take a look at some of the ways to value nature.

We'll do the easy case first. When a good is traded in a market, it is pretty simple to estimate its total financial value: the market price of the good times the quantity of the good sold. And as we saw in Chapter 3, if we have a measure of the supply and demand curves for a good, we can measure the welfare (as

opposed to the financial value) provided by these goods, which is called consumer (or producer) surplus. Some environmental benefits covered in Chapter 4, like timber, rubber, carbon, and wild meat, are typically traded in markets. So it's relatively straightforward to get an estimate of their financial value.

But what about the many benefits from nature for which there are no existing markets? If something isn't traded in markets, does it have zero value? Clearly not; just think of the many ways in which people interact with nature. Robin drives four hours to get to his favorite hiking grounds in British Columbia, so it's important to him, but he doesn't buy or sell hiking grounds. Taylor enjoys the beauty of the lakes and woods in Vermont, but since he can't buy them all up (unless this book sells really, really well), how do we put a value on those experiences? In these and many other cases, there are no formal markets for environmental benefits, but they clearly contribute to our well-being and thus have value.

You can see this distinction in Arden, too. The benefits that Isabella derives from agriculture and logging on her land are relatively straightforward: multiply the price of timber by the quantity of timber harvested, subtract harvesting costs, and these are the net financial benefits that stand to be gained. But Isabella also values less tangible aspects of her property that are not as easy to estimate. To her, the creek that runs through her land is a source of enjoyment because she can swim in it, sit beside it in contemplation, and enjoy the fish flitting in the shallows. How might we estimate this value? And remember from Chapter 2 that while Xavi benefits from Isabella's forested land through the "spillover" of wildlife and the resulting wild meat

from hunting (valued at $200), he also values the aesthetic benefits from his proximity to Isabella's forest at $50.

Estimating values for these kinds of benefits is the purview of a whole subdiscipline in economics called—appropriately—*non-market valuation*. We'll spend the rest of the chapter discussing various non-market valuation methods for quantifying benefits from nature.

> **Non-market valuation** refers to a suite of methods that are used to estimate values—monetary or otherwise—for goods and services that do not pass through established markets. These methods are usually further classified into stated and revealed preference approaches.

Why Value Nature? Why Not?

But wait; let's first make sure we understand *why* we'd want to do this in the first place. Why is it important to value non-market benefits from nature? The strongest answer is from Chapter 2: for better or worse, governments, businesses, and even Brendan's father-in-law often use cost-benefit analyses (CBA) when making a decision. Which option will lead to the greatest net gain (benefits minus costs)? If anything without a monetary value, including environmental benefits, is left out of CBA, then it is given a de facto value of zero. Clearly that's no good, both from an environmental conservation point of view and from the broader perspective of developing public policy that accurately reflects societal values. Of course we can express value in many ways, using many metrics. Conservation proponents often express value in units of rarity or biodiversity. Protected areas are

one way we tend to ensure that value persists, and often we can make decisions based on valuations that don't resort to monetizing the benefits. At the same time, in a world where cost-benefit analyses exist (both formal and informal), monetary valuation of nature gives it a voice within the types of decisions made every day that directly affect the natural world (e.g., the decision to develop a wetland, a coastline, a mountaintop).

There are some additional reasons people undertake valuations. First, it helps call attention to the value of nature in our lives. People may relate better to values expressed in dollars rather than, say, the population size of a species. Monetizing the benefits that we receive from the environment helps to increase society's general understanding of how important nature is to our everyday lives. Second, valuation can help determine the size of incentives necessary to encourage provision of these benefits. Remember the example of Indonesian farmers and erosion control from Chapter 3? Well, what's the value of that reduction in erosion due to conservation agriculture? What's a fair price to pay? This is a critical question for conservation mechanisms called payments for ecosystem services, or PES; we'll get to this concept in Chapter 6, but it is essentially about understanding, then capturing, the non-market value of conservation.

Despite these appealing upsides, the monetary valuation of nature is quite contentious. Many environmentalists are opposed in principle to expressing nature's worth in monetary terms. They argue that once ecosystems are considered on the same economic footing as regular consumer goods such as shoes and toothpaste, we lose sight of the intrinsic value of nature and the moral reasons for sustaining it (Norgaard, Bode, and the Values Reading Group 1998). In economic parlance, monetary valuation will "crowd out" other expressions of our

value for nature.[1] From there, it becomes a slippery slope: if cost-benefit analysis shows that a strip mall would bring more financial benefit than a wetland, then the wetland should make way for superstores and fast-food joints. Others question the accuracy of the techniques that we will describe in this chapter, arguing that they are unable to provide figures that stand up to any kind of scrutiny.

Our own view is that economic values, albeit imperfect, are a useful complement to—but not a substitute for—intrinsic and ethical reasons for protecting nature. We already know that approaching conservation from an economic angle provides insights into the reasons that ecosystems are in trouble, the consequences of their decline, and some clever ways to protect them. Monetary valuation is just one economic tool that helps us in this approach; it doesn't replace the many aesthetic and less tangible reasons for wanting to save ecosystems and species. (In fact, we're having trouble getting Robin to stop bird watching and concentrate on writing what is sure to be a highly lucrative book!) Finally, we should point out that economic valuation is all about looking at trade-offs among various things that are important to people. And although monetary valuation is the most straightforward way to do this, take a look at **Box 5.1** for other important ways in which nature's benefits can be brought into decision contexts without translating values into dollar terms.

---

[1] And it's not just nature here. Paying people money to do something they would have done for free has shown to have a crowding-out effect in a wide range of situations (see Chapter 8). Money sometimes has a negative effect on motivation, which is why your mother probably didn't pay you to clean your room.

BOX 5.1 The Many Faces of Value

It was Oscar Wilde in *The Picture of Dorian Gray* who quipped, "Nowadays people know the price of everything and the value of nothing." Wilde's sarcasm would go a long way in the world of economics, and this line is often the basis for the critique of the monetary valuation of environmental goods and services; i.e., that the price we give such goods and services can never truly reflect their underlying value. In its broadest sense, valuation is simply an attempt to understand the underlying preferences for a certain thing. Value can be measured in many ways. We can imagine a whole suite of metrics that could be used to present the benefits derived from ecosystems, such as "lives saved," "meals provided," "fish available to be caught," etc. In different contexts, different metrics may be appropriate.

For example, when a cyclone struck Orissa, India, in 1999, villages having the most extensive coastal mangroves experienced fewer deaths than those with less mangrove cover (Das and Vincent 2009). The study used "lives saved" and "number of houses destroyed" as two metrics that appropriately demonstrated the value of mangroves. So there is nothing sacrosanct in economics about measuring "value" in monetary terms. However, in many cases we are trying to assess the effects of a decision on a suite of benefits. So the decision to convert a mangrove into a shrimp pond may provide us with improved income and more protein, but it may cost us in terms of lost biodiversity, reduced pole provision, and less effective storm protection.

It is very difficult to compare the costs and benefits of these outcomes without converting them into a common metric. In many (but not all) contexts, the most easily understood

common metric will be a monetary one, like dollars. In some contexts, such as the mangrove case, alternate metrics may be more useful, particularly in cases where the monetary valuation of certain benefits is especially tricky. However, when it can be done in a relatively robust way, monetary values can be extremely useful in making more informed decisions. Remember, our goal is to improve the amount and quality of the information that goes into a decision so that values from nature that are typically ignored are in fact included in the decision-making process. Monetary valuation is one way to ensure this happens.

An Overview of Valuation Techniques

So let's look at some techniques for upsetting many conservation biologists and placing monetary values on ecosystem goods and services. The toolbox of non-market valuation techniques is a mixed bag of methods that each come with a set of assumptions, restrictions, and complexities. Each has champions and detractors, and it's amazing how much economists love to argue about them in specialized journals and darkened conference rooms. Together, though, they give us powerful and quite inventive ways to figure out some of nature's values.

We will start by separating these techniques into *revealed preference* and *stated preference approaches*. Revealed preference techniques are those that observe actual behaviors of people, and so "reveal" their preferences and values. For example, when you go into a clothing store and buy some pants, you "reveal" your preference for those pants in the face of all

other alternative pants. On the other hand, a stated preference approach essentially involves asking people, in sophisticated ways, to "state" their values for particular ecosystem benefits—for example, asking people how much more they would be willing to pay for margarine that is produced in a way that is more friendly to biodiversity than converting old-growth rainforests to oil palm plantations (Bateman et al. 2010).

Revealed preference methods use observed behavior such as market purchases to make inferences on values of goods and services. **Stated preference** methods ask people, sometimes in extremely sophisticated ways, how much they are willing to pay/willing to accept for the gain/loss of a non-marketed good or service.

Both classes of methods have their associated advantages and drawbacks. Revealed preference techniques are generally considered more reliable because they use data on actual behavior. But they tend to rely heavily on sophisticated statistical modeling, with all its associated pitfalls, assumptions, and limitations. Most crucially, they require the existence of a market for an end product to which an environmental good or service contributes. For example, Taylor is hunting for a house in Burlington, Vermont, and proximity to natural amenities such as lakes and forests are as important to him as the number of bedrooms. In this case, we can say that these natural ecosystems are linked to the housing market because they may influence purchasing decisions.

Where such linked markets don't exist, revealed preference methods can't be used, and stated preference techniques are the only way to measure environmental benefits—for example, the value of a pretty landscape where several endemic amphibians live. The upside of stated preference methods is that you get to

design surveys to elicit exactly the values you want to study, instead of relying on markets that may reflect many complex drivers. The overriding limitation of stated preference methods is that they rely on hypothetical rather than actual behavior. So, is what you tell me you'd do what you actually *would* do? To some economists, not being absolutely certain of people's actions is a fatal flaw in stated preference approaches. We will discuss specific pros and cons of each method as we come to them.

I'll Be Watching You: Revealed Preference Methods

OK, let's start with revealed preference and go back again to Arden. The stream that runs through Isabella's property provides her with a number of benefits that she cannot purchase through a market, such as the enjoyment of sitting on its banks and her appreciation of the aquatic life it contains. How might we measure these values? Well, even though there isn't a market for stream enjoyment, there is a market for real estate, and the value of stream enjoyment may be reflected in the value of land. Assuming that many people tend to share her enjoyment of streams, properties that have streams would (all things being equal) fetch higher prices on the market.

This observation—that market goods like land can be decomposed into a set of attributes that define their value—is the basis for what environmental economists call *hedonic pricing*. In the case of Isabella's property, the value of the land is likely determined by a combination of factors, such as its size; the quality of its soil and buildings; its proximity to grocery stores, playgrounds, and restaurants; and perhaps the presence of the stream. To assess the economic value of a stream

using hedonic pricing, you would need to collect information on sales prices of a large number of properties as well as on all property characteristics you think might influence these prices. Then you can use fancy regression techniques to estimate the relative contribution of each characteristic to property value—including environmental characteristics such as the presence of streams. Remember how Taylor wanted a house closer to the forest and lake? He and others with a yearning for nature might very well pay a higher price for such a house, as compared to equivalent houses not near forests and lakes. If this is indeed a general effect, these methods will "reveal" the economic value of lakes and forests to this group of people.

Figure 5.1 shows a simple example from Arden, now zoomed out to show the broader watershed. One can measure, for each parcel, the real estate value and characteristics of the parcel that might affect that value. Using statistics to hold all other factors constant, we can measure the relationship between the parcel's real estate value and its proximity to the stream. For example, parcel 1 is close to the stream and of higher "residual value" (meaning we've indeed controlled for other factors affecting property values) than parcel 2. This imaginary graph shows that properties near the stream have high value, while the further away from the stream one gets, the property value (again, all else being equal) declines exponentially. This is an example of how hedonic pricing can reveal the economic values associated with non-marketed environmental features such as streams.

> **Hedonic pricing** refers to valuation methods that decompose a good or service into the component attributes that define its value. A **production function** approach is analogous and is the term used in recent ecosystem services literature.

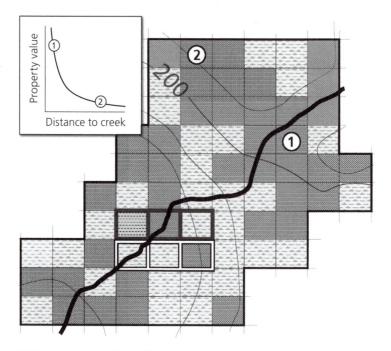

FIGURE 5.1 Arden from Chapter 4, zoomed out to show the broader watershed. A creek runs through the middle of the basin, with Isabella's and Xavi's properties near the southwest corner. The graph shows the relationship between property values and proximity to the creek (holding other factors constant via hedonic pricing), with parcels 1 and 2 as examples.

There are numerous examples of the contribution of nature to real estate prices in various parts of the world. In Castellón, Spain, sale prices of urban houses declined by 1,800 euros for every 100 meters further away from green space that the house was located (Morancho 2003). In Grand Rapids, Michigan, sale

prices of undeveloped lots were between $5,800 and $8,400 higher for lots adjacent to the border of a tract of protected forest (Thorsnes 2002). And in the Netherlands, houses that faced water bodies commanded price premiums of up to 28% compared to similar houses without such an amenity (Luttik 2000). Clearly, the benefits from at least certain types of natural environments can be captured by using the hedonic pricing approach. Interestingly, it's not just the obvious proximity to magnificent views and water bodies that counts, nor even the trend for wealthier people to buy expensive houses in wilderness near protected areas (Radeloff et al. 2011): wealthier neighborhoods in Phoenix, Arizona, have a higher diversity of plants, suggesting that properties higher in biodiversity are more valuable than those with lower diversity levels (Hope et al. 2003).

Though *hedonic pricing* is the term used in the economics literature and has a long history, especially as applied to real estate, the nascent ecosystem services literature tends to use the term *production function* to refer to valuation approaches that decompose a marketed good or service into the characteristics (including some measure of the environment) that define its value. You will likely come across both if you continue reading articles in this field, and you can treat them essentially as synonymous. Personally, we prefer the term *production function*; it seems more intuitive to us to think about how the value of any particular good or service is "produced" by a variety of factors. So we'll go with this term for the rest of the chapter.

Another example of the production function approach is in the estimation of the value of crop pollination by bees. Crops are sold in markets, so we have a price for the final product. But how much of that value is attributable to the bees that pollinate those crops? What is that pollination service worth to the farmer?

This is exactly the question that Taylor and colleagues asked in Costa Rica using a production function for—you guessed it—coffee (Ricketts et al. 2004). Coffee plants, like about two-thirds of global crops, produce more and better crops when they are pollinated by bees and other animals. Taylor hypothesized that coffee fields closer to forests would enjoy higher pollination services from wild bees than similar fields farther away. Bees would emerge from forests to pollinate nearby coffee when it flowered, and yields would be improved as a result. He tested this hypothesis by studying coffee plants at increasing distances from forest. By carefully selecting coffee plants within a single large farm, he kept most of the other important factors constant, such as coffee variety, soils, fertilizer input, labor, and management style. So what he wound up with was a very simple production function, relating coffee yield to forest proximity.

The results showed that plants close to forest patches received many more visits by many more species of bees. And coffee yields responded; within about a kilometer of the nearest forest, plants produced about 20% more coffee, and of higher quality, than those farther away. Armed with this simple production function, Taylor calculated that if two specific forest patches were destroyed, much of the farm would wind up more isolated from tropical forest than it had been, and the owners would lose coffee production worth $60,000 per year in net revenue. This number is a good estimate of the economic value of those forest patches to crop pollination on that farm. The key to getting such an estimate was fitting the production function.

Moving on, travel cost methods are another set of revealed preference techniques. These methods are often used in studies of tourism, when people travel to a particular place to enjoy the natural features or beauty of an area. In these cases, we can

estimate the recreational value of a site or set of sites. The basic intuition is that preferred places will have more visitors, and from farther away, than less preferred sites.

Travel cost methods are those that estimate economic values (typically for some form of recreation) based on the assumption that traveling farther to get somewhere is, all else being equal, an indication that the destination site is more highly valued than a site that would take less time to get to.

There are two flavors of travel cost models. In its simplest format, the hedonic travel cost method uses information on the number of visits to a particular site, in combination with distance traveled by each visitor to the site, to estimate how valuable sites are to people. Given information on how many visitors have visited a particular site and from how far they have come, basic methods like the per-kilometer cost of driving are used to convert distance traveled into a dollar estimate of value. A demand curve for the site can be constructed by plotting the number of visits against distance traveled (cost), and from this the consumer surplus can be calculated to get an estimate of the welfare gained from recreation. Building out from this, if we have a set of sites with observable characteristics such as infrastructure levels and environmental amenities, we can use regression methods to quantify the importance of factors other than costs to visits. The other variant of the travel cost method treats the decision to visit or not as a function of the observable characteristics of the site and the visitor; this "random utility" approach to measurement has somewhat greater data requirements, as we shall see next.

For example, every year tens of thousands of people visit beautiful Jasper National Park in the Canadian Rocky Mountains to enjoy its spectacular scenery and abundant wildlife.

But what drives visitation to different parts of the park, and can we quantify how much these visits are worth to people? A study addressed this question by collating information from over 1,600 backcountry permits issued to hikers hailing from Canada's four majestic western provinces (Englin, McDonald, and Moeltner 2006). Information on each permit included the park trail to be hiked, the home address postal code, the group size, and the number of days spent in the park. Multiplying distance from the home address to the trailhead by a flat dollar value per kilometer driven resulted in a cost estimate for each trip. The authors modeled the decision to visit any particular trailhead as a function of cost and also observable characteristics of the environment that trails pass through. In addition to the expected finding that the likelihood of visiting a trail is lower the more costly it is (i.e., the further away a visitor is coming from), they found a strong positive effect on visitation of the amount of old-growth (ancient) forest that a trail passed through. They used their models not only to estimate the aggregate per-trip consumer surplus (about CAD $500) but also to show that welfare changes from increased encounters with old-growth forest could be as high as several hundred dollars. So travel cost models can help us understand not only the overall recreational value of tourist sites but also the effect of particular attributes on measures of value—in this case, there is a clear and significant value of old-growth forests to hikers.

We humans often don't realize how much we have until we lose it, and this applies as much to ecosystem services as to people we used to be in a relationship with. Think of the lost wetlands of Louisiana that might have absorbed the waves and storm surges of 2005's Hurricane Katrina, which could have prevented much of the damage to the city of New Orleans. The

replacement cost method uses the value of a similar alternative good or service to estimate the value of something that has been lost. Think again of Xavi in Arden. He benefits from Isabella's land in several different ways, including the wildlife spillover and the resulting wild meat available to him and his family. Since Xavi doesn't sell that meat, we cannot use market prices to estimate its value. However, if Xavi lost access to this wild meat and had to replace it with beef bought at the local market, it would cost him $200. Voilà! Replacement cost.

Assessing the *damage costs avoided* is a similar method that assumes the value of a good or service that would be lost through a change in the amount or quality of an ecosystem is a measure of the value of that ecosystem change. Essentially, the physical relationship between a change in environmental quantity and quality and some physical measure of damage (amount of crop lost, days unable to work, flooding of real estate, etc.) is quantified, and then per-unit values of the damage are used to calculate the cost that would be avoided by the environmental conditions or change. For example, the fact that your house did not get flooded because the local wetland absorbed the water means that some approximation of (at least one) value of the wetland is the value of the damage you didn't experience. Thanks, local wetland!

Ask and You Shall Receive (Data): Stated Preference Methods

Certain goods or services produced by the environment neither are sold via markets nor contribute to the production of any marketed goods. For example, existence values are the values that people hold for something simply knowing that it exists, regardless of whether they ever plan to see or use it. For these types of benefits, there is no way to use revealed preference approaches to estimate their value. You simply have to ask people. It may sound flaky, but many people could name a few special species or places that they highly value but yet never expect to see or use. These values are real and important; hence, economists have developed some surprisingly sophisticated stated preference techniques to quantify them.

Stated preference methods ask people, sometimes in extremely sophisticated ways, how much they are willing to pay/willing to accept for the gain/loss of a non-marketed good or service.

Two main classes of survey instruments measure stated preferences: contingent valuation and choice experiments. Contingent valuation surveys typically describe a particular environmental policy or decision and ask respondents how much they would be willing to pay to ensure the preservation of a particular species, area of tropical forest, and so on (or less commonly, how much they would be willing to accept to allow the loss of these things).

Contingent valuation methods use surveys to ask people how much they would be willing to pay for a non-marketed ecosystem good or service. **Choice experiments** decompose goods, services, and scenarios into attributes and ask people to make a choice based on the values of these attributes. Both approaches allow estimates of peoples' values for non-marketed goods and services.

Questions can be open ended, based on a predetermined scale, or posited as a yes/no answer to a stated dollar value.

Let's take a look back at Arden and see how factoring in these values might affect Isabella's land-use decisions (**Figure 5.2**). Converting the eastern forest parcel would not only emit carbon into the atmosphere but also eliminate the habitat of many forest birds whose beautiful songs both Xavi and Isabella enjoy listening to. Although the stream remains now that the land is a coffee plantation, sitting beside it is a little less enjoyable without the accompaniment of birdsong. What is the lost enjoyment from this environmental amenity worth? We don't have markets to reveal these preferences, so we ask Isabella and her neighbors via a contingent valuation survey. Remember from Chapter 2 that Xavi already stated a general aesthetic value at $50. From this survey we find that both Isabella and her neighbors would be willing to pay an additional $100 per year to ensure that forest birds could persist with their singing. This addition to the CBA makes an even stronger case than in Chapter 4, at both private and social levels, for conserving the forest parcel rather than converting it.

How about a real-world example? In Iran, the remaining forests in the north of the country sheltered the world's largest subspecies of leopard (*Panthera pardus saxicola*) and were also the haunt of the Caspian tiger (*Panthera tigris virgate*) until its extinction over 50 years ago. Given plans to reintroduce the Caspian tiger from its genetically indistinguishable cousin, the Amur (Siberian) tiger (*Panthera tigris altaica*), the biggest cat on Earth may again stalk these forests sometime soon (Driscoll et al. 2012). Regardless, these forests' conservation value in a largely arid country is surely quite high . . . but how can such a value be quantified? This question was tackled using

□ Isabella	□ Xavi	▨ Forest	⊡ Sparse forest	≋ Agriculture

BENEFITS	**COSTS**		
Private benefits	*Private costs*		
Timber revenue........$250	Timber harvest........$200		
Agriculture revenue..$250	Crop planting, harvest...................$200		
	Lost carbon payment..$200		
Social benefits	Lost birdsong..........$100		
	Social costs		
	Loss of hunting and aesthestics for Xavi..$125		
	Social costs of carbon..................$1,600		
	Lost birdsong..........$100		
		Difference	Ratio
TOTAL......................$500	**TOTAL....................$2,525**	($2,025)	0.2
Private total.............$500	Private total.............$700	($200)	0.7
Social total...................$0	Social total............$1,825	($1,825)	0.0

FIGURE 5.2 Isabella's and Xavi's properties in Arden, adding consideration of the aesthetic values of listening to birds along an undisturbed creek. These considerations add private and social costs to converting the forest parcel, and the resulting cost-benefit analysis indicates it should be conserved.

a contingent valuation approach, asking respondents in Iran whether they would be willing to pay a certain level of special tax to fund a new management program that would improve the condition of the country's northern forests (Amirnejad et al. 2006). Their sample of 950 people showed that there was a mean annual willingness to pay of USD $30.12 for such improved management, and that younger, more highly educated, and wealthier individuals had the highest WTP values. If the sample is representative of households across the country, it would mean that people in Iran hold significant values for the persistence of their forests, to the tune of USD $200 per hectare, or about $375 million in aggregate for annual management. Having a solid figure on the economic value of standing forests in Iran's north could help sway policy decisions toward conservation and away from continued destruction of these forests.

This is one example we pulled from hundreds of contingent valuation studies that have assessed non-use values for ecosystem goods and services, and it actually appears to be at the high end. A review of non-use values for the world's forests suggests that values of $2 to $45 per hectare are typical (Pearce 2001). Other studies of developed-country citizens (Horton et al. 2003, Kramer and Mercer 1997) suggest that stated willingness-to-pay values from contingent valuation surveys would be enough to generate on the order of 1 to 3 billion dollars for rainforest conservation, if such values are applicable to the entire country's population. And a review of contingent valuation studies on a variety of threatened species shows a wide range of values, from USD $8.32 per household per year to avoid the loss of the striped shiner (*Luxilus chrysocephalus*) to $100 or more for the persistence of various game fish (Richardson and Loomis 2009).

Though evidently quite variable, such values are not just in the realm of the theoretical and academic. A blue-ribbon panel convened by the National Oceanic and Atmospheric Administration in the United States, which included two Nobel Prize winners, concluded that contingent valuation studies are acceptable sources of monetary values for use in cost-benefit analysis and decision making. And since the Exxon Valdez oil spill off the coast of Alaska in 1989, contingent valuation studies have provided estimates of the losses that at-fault companies must compensate as a result of environmental damages. Current work in this regard includes estimating the damages, including lost existence values, from the 2010 Deepwater Horizon oil spill in the Gulf of Mexico.[2]

Despite the many academic studies and practical uses of contingent valuation, a whole slew of design rules must be followed if the surveys are to produce reliable value estimates. Some of the key points are as follows. First, the value elicited is contingent on the hypothetical market described in the survey. Respondents must be sufficiently informed of the context and background of whatever policy decision is being considered. Second, both the scenario and the payment vehicle (jargon for how you are going to pay, such as higher prices, a tax credit, etc.) must be realistic and believable, so we are as confident as possible that values expressed by the respondents indicate what they would do in real life. Third, sampling issues (what

[2] Saying that everyone does not agree with this approach to value damages from environmental disasters is a bit like saying a wildebeest finds it annoying to have a lion's teeth sunk into its neck. The academic debate here is heated and actually kind of fun to follow.

is the appropriate sampling design—randomized, stratified random?) and survey presentation (what is the right balance of text, images, and background information?) must be carefully considered given the target population. Finally, pre-tests, post-survey follow-up questions, and additional queries regarding attitudes and demographics can help ensure solid interpretation of the results.

Seems relatively straightforward . . . but many critics in fact argue that contingent valuation studies cannot provide robust estimates of values. Doubters insist that such studies simply measure the "warm glow" that many people get from saying they are willing to pay for something good for the world and not for the actual ecosystem good or species that is supposedly being measured. The hypothetical nature of the survey may also result in embedding, in which respondents' willingness-to-pay values are insensitive to the quantity of the good being valued, which may indicate that they are not taking account of the fact that their ability to pay for the goods is constrained by their own household income (in other words, the hypothetical context is not reflecting what would actually happen in real life). Additionally, *anchoring* refers to the fact that surveys having a particular choice of dollar values may bias their respondents toward an "anchor" value in the presented set that may be quite different from their actual WTP.

Choice experiments are a more recent stated preference technique, developed to address many of the drawbacks of contingent valuation. Choice experiments were originally developed in marketing research as a means of assessing consumer preference for as-yet undeveloped consumer products. Choice experiments use the same logic as production functions, assuming that a product (or more broadly, a choice that

consumers will make) can be decomposed into component attributes that define its value. To get at the relative weighting of the different attributes, you ask people to make a series of choices—selecting the alternative based on the particular combination of attribute levels that they most prefer. These choices are carefully designed to vary the component attributes independently, so that at the end you can estimate the importance of each.

Another example that we've already brought up may make things clearer. Robin and Vic's study of birds in Uganda (not birds again!) used a choice experiment to estimate people's preferences for visiting forest reserves based on a variety of attributes, including the number of bird species they were likely to see. **Figure 5.3** on p. 108 shows one of 16 scenarios that visitors to Uganda evaluated as part of this survey. The visitors were asked which of the three parks, if any, they would choose to visit, given the attributes laid out in the scenarios. Robin and Vic's 15 other scenarios each had different attribute levels for each choice of park. With enough replication in each category and a variety of variable combinations, they were able to observe how these attribute levels affected people's decisions, and they constructed a function that weighted the importance of each attribute to the respondents' choice of park to visit. In some ways, what they found was not surprising: the dominant element that predicted whether people would visit a park was the chance of seeing large wildlife. And the longer the travel time to a park, the less willing people would be to visit it. These results may seem obvious, but given they worked in the way we expected them to, it gives us confidence that the scenarios were being read and understood by people who were responding in logically consistent ways to the choices they were presented. A more surprising result was

1) If the following nature parks were the ONLY THREE choices available when you were making your decision on where to visit on your next trip out of Kampala, which one would you have picked?

Features of nature parks	Mabira Forest Reserve	Budongo Forest Reserve	Kibale National Park	
Travel time from Kampala	1 hour	5 hours	6 hours	I would **NOT** visit any of these protected areas on my next trip
Entrance fee (U.S. $)	25	25	25	
Destination part of package or tour?	No	Yes	Yes	
Lodging facilities	Luxury lodge	None	Luxury lodge	
Landscape features	Primary forest	Primary forest, secondary forest, and agriculture	Primary and secondary forest	
Bird species you may see	40	80	60	
Chance of seeing large wildlife	Very slim chance	Very good chance	Very good chance	
Please Tick ONE Box Only	☐	☐	☐	☐

FIGURE 5.3 One of the 16 scenarios that visitors to Uganda evaluated as part of Naidoo and Adamowicz's (2005) choice experiment. The visitors were faced with choosing one of three nature parks to visit, as indicated by columns 2–4; column 5 offers the choice of visiting none of the three options. The first column gives the names of the attributes that are assumed to fully define each alternative, and the attribute levels are given below each listed park.

just how much importance respondents attached to the number of bird species that they were likely to see; this was a strongly significant variable in people's choice of park to visit, and it was on the basis of this response that subsequent calculations revealed that ecotourism alone could offset the opportunity costs of conserving almost 90% of Mabira's forest birds (Naidoo and Adamowicz 2005).

Transferring the Benefit

In some cases, none of the valuation techniques described in this chapter are feasible. Data may not exist, the scales may be too big for survey techniques, or estimates may be needed quickly so there isn't time to do careful analyses. A common fallback solution is to "transfer" published estimates of value (e.g., the per-hectare value of a wetland) to an area of interest where no such studies have taken place. Perhaps the most prominent example of this approach was a paper by Costanza et al. (1997), in which the value of *all* of the world's goods and services was estimated by applying values derived from a set of studies for different biome types to every hectare of this biome type across the entire Earth.

This approach has several problems. It is very difficult to match the context of the original valuation to that of the place of interest. How can you be sure that all the ecological and socioeconomic characteristics of the situation are the same between any two areas? A failure to consider the characteristics of the new sites to which values are being transferred can produce nonsensical results. For example, the total economic value of the crop services that bats provide in the United States was recently estimated by extrapolating from research in a cotton-dominated

agricultural area in south-central Texas (Boyles et al. 2011). This benefit transfer produced results that suggested the economic value of pest control services by bats was up to 50% of the total crop value in a number of states whose crops were not actually affected by the bat-regulated pests in the first place (Fisher and Naidoo 2011)!

An improvement on simply transferring unit values across study contexts is to instead transfer the function from which values were derived. For example, if you undertake a choice experiment to elicit values of wetlands in southern England and get an average estimate of, say, $50 per household per hectare, a benefits transfer approach would suggest that you could assign $50 as the value for a hectare of wetland held by households elsewhere. That might make sense if "elsewhere" was in the Midlands of England, but what if it was in Canada? or Kenya? or Djibouti? Those contexts are unlikely to closely match the conditions of the original study in southern England. However, if the $50 value was a function of respondents' household income and proximity to other natural areas, more accurate results might be obtained by using a valuation *function* rather than the average value itself. Of course, this method requires slightly more work because data for the variables that make up the function will need to be collected for the new location. This may nevertheless represent a good trade-off, getting reasonable estimates at reasonable cost rather than the extremes of a completely new field study, or else transferring the average value.

We've tried in this chapter to give you a sense of some of the main non-market valuation techniques that you're likely to run across in thinking about your own work. In the spirit of a primer, we have naturally been unable to cover these as thoroughly as we might have liked. Read through the deeper

environmental valuation literature and you'll be well on your way to calculating the economic values of nature—but then what? Even if we demonstrate how valuable nature is, in some cases we still need a way to *capture* these values in situations where mechanisms like markets don't exist. This is the topic of the next chapter, so read on!

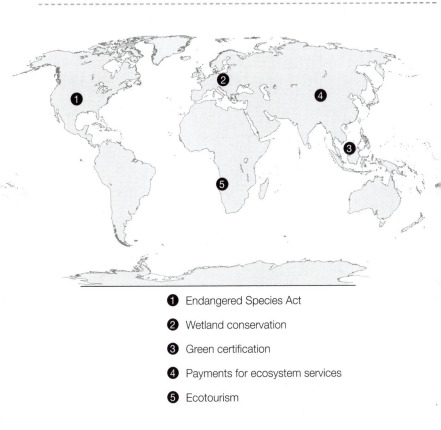

1 Endangered Species Act

2 Wetland conservation

3 Green certification

4 Payments for ecosystem services

5 Ecotourism

Institutions

Setting the Rules for Conservation

In Chapter 5 we discussed several methods of estimating the economic value of various types of non-market benefits. Let's hearken—yes, hearken—back to the reason we wanted to estimate these values in the first place. Ecosystems and the biodiversity they contain provide us (as individuals and as a society) with a whole suite of goods and services for which there is often no explicit acknowledgment of value. When a factory burns down, we immediately hear about the loss of property, loss of production, and loss of jobs, all of which have a clear financial value that we can calculate. However, when a forest burns down, we do not immediately get a ledger showing how much carbon has been lost, how the hydrological cycle of the area will change, or how many species disappeared. If there were a clear indication of how much economic value was lost in this instance (remember that it doesn't have to be expressed in dollars—economic value

is about utility), perhaps our policies, management, and interactions with the natural world would change. Packed into our factory scenario are two important concepts that strongly link the work we do in conservation with the topic of this chapter. Those concepts are *information failure* and *market failure.*

In broad strokes, information failure occurs when there is not perfect information among people in an exchange, meaning there are things not known about the good or service being exchanged (e.g., hidden costs, extra benefits). In the world of conservation, this lack of information is ubiquitous. It is very hard to make management and policy decisions at meaningful scales that reflect all of the values for conservation because we simply do not have all of the information. How will a coral reef system react to increased tourism or fishing pressure? How will a hydroelectric dam be affected by deforestation 200 kilometers upstream? Sometimes the essential information is only known by folks on one side of the exchange; this is a particular form of information failure, called asymmetric information. So if a conservation NGO (non-governmental organization) offers you money for your land because of its biodiversity value, you might sell at a given price (perhaps for something at or above your opportunity cost). But if the NGO knew that there was valuable oil under your property but you did not, then they had information you didn't—there's

the asymmetry. If you knew that the oil was there, you might sell at a much higher price.[1]

Market failure, on the other hand, is related to the public or common-pool nature of the goods and services that nature provides. Remember from Chapter 4 that the market is pretty good at providing us with shoes, pants, chickens, and goats but not so good at allocating the "correct" amount of biodiversity, carbon storage, and sediment retention. With things like biodiversity and many ecosystem services, the market fails to provide the good at all or fails to provide the right amount—market failure.

So, we don't have all of the information nor the markets we need for providing the efficient amount of a given ecosystem service. But we do know nature has value. And that is where institutions come in.

As the late David Pearce wrote many times, showing that the environment has value, the topic of our whole previous chapter, is only the "demonstration" phase. To truly move beyond simply demonstrating that nature has value, we need to develop policies and management tools to "capture" these values. In this context, that means ensuring that the benefits that ecosystems and biodiversity provide to people are understandable and important to decision makers. If these previously uncaptured benefits can be made tangible, then the incentive for choosing conservation increases. So, how can we capture these values, in light of information and market failure? Well, we do this through institutions.

---

[1] OK, so we made two assumptions here: (1) that you were acting like a monetary profit maximizer, which we are sure you are not—why else would you be in conservation?—and (2) that the NGO was a sneaky one. (We made the latter assumption simply because we were tired of blaming logging companies, palm oil companies, etc. We might get back to that later.)

What Are Institutions?

Institutions? We don't know about you, but when someone says "institution" we typically think of some huge marble building with columns in front, filled with filing cabinets, paperwork, TPS reports, grumpy people, and bad coffee. But when Brendan puts his economics hat on, he remembers that institutions, to an economist, represent something different. To an economist, the word *institution* refers to a set of formal or informal rules that govern human behavior. That may seem like a pretty broad concept, so let's jump back into our manufactured dialogue among coauthors to get a bit more concrete.

> In economics, an **institution** refers to a set of formal or informal rules that govern behavior. With respect to conservation, things like marine protected areas, endangered species legislation, biodiversity offsets, and carbon markets would all be considered institutions.

SETTING: *Someplace.* Cast*: Robin and Taylor.*

TAYLOR: *I just read your definition of* institutions. *So in economics,* institutions *is a pretty broad concept, huh?*

ROBIN: *Yep.*

TAYLOR: *So does that mean there are many different kinds and forms of institutions?*

ROBIN: *You betcha.*

TAYLOR: *So by that definition, a biodiversity offset that governs behavior of, say, a mining company around biodiversity protection is an institution?*

ROBIN: *Yes.*

TAYLOR: *How about a law regarding fishing rights in Indonesia?*

ROBIN: *Yep, that's an institution.*
TAYLOR: *What about a tax on pollution that affects salmon streams?*
ROBIN: *That's an institution, too.*
TAYLOR: *What about a goat?*
ROBIN: *Not an institution.*
TAYLOR: *OK. Got it.*

So by *institutions* we mean a whole suite of rules that dictate how people within societies interact with one another. In the context of conservation, institutions provide some structure to help us govern and manage our natural world. But let's make this concept a little more manageable. One typical categorization of institutional arrangements is to divide them into "market-based" and "regulatory." This dichotomy is somewhat artificial because market-based mechanisms often require strong regulatory environments. Also, it leaves out a whole range of informal arrangements that people and communities use to manage their natural resources. That said, it is a useful construct for better understanding institutions with respect to conservation.

Regulation as an Institutional Arrangement

Regulations govern our behavior in many ways. Some governments outlaw logging, some require pollution control devices on cars, and some set minimum standards for water quality. In short, regulatory institutions help manage, sustain, and/or provide goods and services that would otherwise be undersupplied by private markets. Highways, national defense,

health care, and public transportation are examples of services that are subsidized and institutionalized by some governments, thereby correcting the failure of the market to provide these things in a way that is desirable.

Protected areas—the bread and butter of conservation activities over the past century—are regulatory institutions. They are a way of society saying, "Hey, we're not going to let anything happen to this patch of land or sea, and so the biodiversity within it will be safe." Now of course there are detractors to this model and lots of talk about "empty forests" and "paper parks." However, some recent analysis by Jonas Geldmann and colleagues (2013) has shown that despite all of the problems we have governing protected areas, they are, in general, quite an effective institutional arrangement for protecting habitat from conversion. What is less clear from this work is how that protection affects populations of species within protected areas.

Moving over to the marine realm, let's consider the Marine Mammal Protection Act in the United States. Established in 1972, the main goal of this institution was to maintain marine mammal stocks within US waters. And although we don't have good data for a lot of these stocks, about 60% of populations we do have data for are increasing, about 20% are stable, and about 20% are decreasing (Roman et al. 2013). In general, this legislation has created a pretty effective institution for protecting marine mammals.

We have painted a fairly rosy picture, huh? But how about this one: the US Endangered Species Act was established in 1973, when the avian symbol of the United States (the bald eagle) was about to disappear. This landmark legislation has had decades of praise, as well as decades of voices raised against its

effectiveness, and even its constitutionality. Using a technique known as statistical matching, Paul Ferraro and colleagues (2007) were able to control for the biases in the way species actually get listed for protection and truly test the effectiveness of the act. What they found is that, on average, listing a species as "endangered" actually had a detrimental effect on that species' populations.[2] However, if the listing of a species was combined with substantial government funding for the protection of that species, then the status of that species was likely to improve.

OK, so institutions don't automatically solve everything. There are some trials, some tribulations, and even some counterintuitive outcomes. Can you remember the situation in Ghana all the way back in Chapter 2? The link between fisheries and bushmeat harvesting in Ghana only arises because of the particular institutional arrangement in place. Ghana allows international fishing fleets into its coastal waters to fish, resulting in overfishing that has reduced the supply of fish to coastal villagers and hence caused them to hunt more bushmeat for food in certain years. This particular institutional arrangement (the allowance of foreign fishers in Ghanaian waters) allows the second-order links between bushmeat harvest and fisheries harvest to emerge. Under a different set of institutional rules, such a link might not be as strong.

Let's look at one more example. This one has to do with China's Sloping Land Conversion Program, which was devised to reduce erosion in the headwaters of the Yangtze and Yellow

---

[2] For an interesting discussion of why this might be—as well as the reason for the former popularity of the "I Eat Red-Cockaded Woodpeckers" bumper sticker in the southern United States, read Andrew Balmford's *Wild Hope,* Chapter 3.

Rivers by restoring millions of hectares of cropland on steep slopes into forest or grasslands. Under this program, landowners could sign up and receive payments in grain for converting their plots on slopes greater than 15 degrees. So here's an institution that people "opt in" to be a part of. They get "paid" (in grain) as compensation for taking some of their land out of production. So the institution corrects the failure of the market (or livelihood strategy) to provide the right amount of a public good (erosion control). This pretty cool approach is an example of a payment for ecosystem services (PES) that we've been talking about in Arden. And it seems to be delivering. Liu et al. (2008) and Li et al. (2011) show that the program has been responsible for the retirement of millions of hectares of cropland on steep slopes, has reduced runoff and soil erosion, and has provided income benefits for middle- and low-income households.

This last example is somewhat different from the others, right? There's a policy in there, but also a payment, and it doesn't look like strict regulation. Remember what we said about a false dichotomy? We'll talk more about this in the next section.

Market-Based Institutional Arrangements

Rather than using regulation to govern behavior, as in the examples above, a whole range of conservation institutions use the power of markets to meet desired ends. These include taxes and user fees to curb undesirable behaviors as well as payments and subsidies to encourage desirable behaviors. Such approaches use price incentives and the efficiency of the market to connect buyers and sellers to achieve something.

For example, in addition to producing the British Columbia Grizzly Bear Conservation Strategy, the government of British Columbia[3] has levied a tax on the burning of fossil fuels. This amounts to a tax of about 7 cents per liter of gasoline. If you are a car owner in British Columbia, you basically have two choices: drive less because you don't want to pay the extra tariff on gas, or pay the tax and drive the same amount as you always have. If you choose the first option, then the price disincentive (extra cost) passed on to you via the market has achieved the goal of reducing the amount of fossil fuels being burned. If you choose the second option, then the price incentive wasn't strong enough to change your behavior, but the government did raise money—perhaps to invest in solar infrastructure (or grizzly bears).

On the flip side of a tax is a subsidy. A subsidy is simply an institution that uses a reward to incentivize good behavior. Several countries across Europe have subsidies for wetland protection, basically another payment

> A **subsidy** is a reward or payment to encourage the uptake of a policy, program, or institution; or to aid in overcoming any cost burdens a policy, program, or institution may introduce.

for the protection of an ecosystem (and the provision of its services!). In the United States, there is a policy on the books informally known as the *No Net Loss* rule for wetlands. Under this policy the total area of wetlands in the country is not allowed to go down in order to maintain the myriad services they provide. This kind of sounds like a regulatory approach, right? Well, sure. But under this policy some states help to operate wetlands banks that work like market instruments. So, say the

[3] That's in Canada, eh? One of the authors loves Canada.

Not-Evil-But-Not-So-Good-Either Company (NEBNSGE Co.) wants to convert a coastal wetland into a new production facility for their ethically neutral product. Under No Net Loss, the company must compensate for the wetland's loss via the restoration of an "equivalent" wetland offsite. This restoration can take the form of a purchase of "credits" from wetlands that have been conserved by others and certified by the US Army Corps of Engineers according to the restored wetland's area and functional significance.

Of course, there are many reasons why it would be nearly impossible to create or restore an "equivalent" wetland. How could we ensure that it functions the same way, with the same species composition, while delivering a similar level of benefits to the same people? These are real problems with this institution that need to be resolved if No Net Loss is going to mean more than the creation of a biologically meaningless patch of soggy land. A wetland is not a wetland is not a wetland.

Such wetland credit and banking systems in the United States represent a market approach known as cap and trade, where some physical "cap" is set—in this case the cap is net zero; i.e., the total area of wetlands in the United States will not change. However, how this balance is achieved can be left up to a market of buying, selling, and trading wetland areas. So in our example above, NEBNSGE Co. can pay another landowner for their wetland credit (or restoration project) and then convert the wetland that just so happens to be sitting in the same spot as their future production plant. Cap-and-trade policy has been used to effectively reduce the sulfur dioxide emissions that cause acid rain, and it is currently on the table as a potential instrument for mitigating CO_2 emissions. Similar offset schemes are in development for biodiversity, but they are still

in the nascent stages (see Gardner et al. 2013 for if, when, and how biodiversity offsets can work for conservation).

The above market-based mechanisms involve a relatively small number of participants. Are there any schemes that attempt to harness a larger portion of the overall consumer market? Well, sure. A growing number of internationally traded products now have certification standards for environmentally friendly production practices. Examples you may have come across are shade-grown (and therefore potentially bird-friendly) coffee, Forest Stewardship Council (FSC) certified timber production, and Marine Stewardship Council (MSC) certified fish. The theory here is that a certain segment of the consumer market is willing to pay a price premium to be assured that the products they buy are produced in an environmentally friendly manner. The price premium must be sufficiently high to cover the increased production costs associated with environmentally friendly practices. If it does, then we can overcome market failure and incentivize producers to adopt sustainable practices.

Take FSC-certified timber. For wood to be certified, timber companies must be inspected and must convince the FSC auditor that their harvesting practices allow, among other criteria, the persistence of ecological functions and integrity of the forest. They must also protect areas identified as "high value" for conservation. Both the opportunity costs of leaving trees standing and the cost of monitoring and verification need to be offset by the increased price paid by at least some consumers for the resulting certified timber. The principle is similar for all other certification schemes.

For a company evaluating whether it makes financial sense to get involved in the certification business, the costs of doing so (forgone product revenue and inspection fees) are relatively easy to

calculate; but how can the benefits be assessed? Companies need to assess whether the market's willingness to pay for a more environmentally friendly product is sufficient to offset the associated increased costs. Market research techniques, such as the choice experiment methodology introduced in the last chapter, can play a key role in a company's determination of whether the overall size and significance of the "niche" market is enough to warrant the production of an environmentally friendly product.

Ian Bateman and colleagues (2010) used choice experiments to assess the potential of the market for biodiversity-friendly palm oil.[4] Palm oil production in Southeast Asia, especially in Malaysia and Indonesia, is booming to satisfy demand for biofuels, which has resulted in a dramatic reduction in the region's megadiverse primary forests and has increased concern from conservationists on how to protect what remains. Environmental groups have spurred the formation of a round table on sustainable palm oil production, to agree on production standards and perhaps certification of wildlife-friendly palm oil. Bateman et al. investigated the flip side: what is the size of the market for biodiversity-friendly (in their specific case, tiger-friendly) palm oil? They used a choice experiment to assess demand and willingness to pay for a tiger-friendly margarine product versus a conventional one and found that consumers are on average willing to pay between one-third and one-half more for high-quality, tiger-friendly margarine. Follow-up work from the same group suggests that this premium

---

[4] Anyone who has ever walked through an oil palm plantation will wonder if it is even possible to produce biodiversity-friendly palm oil, but one such action that land managers can take is simply not planting oil palms in low-productivity areas that abut primary or secondary forests.

is enough to offset the opportunity costs associated with more biodiversity-friendly palm oil production. However, that result is only for production areas where the cost-benefit ratio is fairly high. In other words, people will pay more for "greener" margarine—and that extra profit will offset the costs of "greener" production—but only on lands that aren't that profitable in the first place.

We've said it before, but this example shows why economics is so important to conservation. By using an "institution" (green product), we can align the incentives of the private company (to make money) with the goal of conservation (to save stuff).[5]

We've just touched the surface of these market-based instruments. There are a slew of them we didn't cover here, including individual transferable quotas (ITQs) in fisheries, ecotourism, and debt-for-nature swaps. Our broader point is that by utilizing the market (e.g., a cap on wetland loss, certified product differentiation) we can more efficiently supply a level of public good or common-pool resources that would not have been supplied without the institution itself. Now the astute student (which we are sure you are) will say, "Hey, wait a minute, all of those market-based institutions required regulatory settings and compliance!" That's right, our cap-and-trade work needs a cap policy; our certified markets need some quality control and consumer protection; our tax and subsidy need government oversight and the ability to sanction noncompliance. So yes,

---

[5] We are not endorsing one of the worst drivers of change for biodiversity in the past three decades (i.e., the expansion of oil palm plantations across Southeast Asia), but rather we are highlighting another instance where economics sheds light on the possibility to manage landscapes in a more desirable way with respect to conservation.

markets need regulation, too. For now let's explain a few of the operational differences between using a regulatory (also known as command-and-control) approach and a market-based one.

Regulation or Markets, Regulation and Markets, or Neither

So we have these types of institutions that can help us to achieve more desirable levels of conservation. We hope you've picked up on why in some cases a market-based approach is a good idea while in others the regulatory approach is preferable. But let's break it down using the example of conserving wetlands and think about it with regard to a few important criteria upon which to judge which is the better approach. In **Figure 6.1** we consider two approaches to wetland conservation in a given country. The first is simply a law that forbids the conversion of

	Cost-effectiveness	Fairness	Incentivizes innovation	Ecological outcomes
Regulation: No wetland loss	☐	✓	☐	✓
Market: No net loss with credits	✓	☐	✓	☐

FIGURE 6.1 Scoring regulatory and market-based approaches for wetland conservation. Check marks indicate the preferred institution for each of four criteria. (See Burger et al. 2009 for a carbon emissions example.)

any wetland. Let's assume that the fines are severe and the government has an excellent ability to monitor for compliance. The second approach is to legislate the No Net Loss rule and allow for a market where stakeholders can buy and trade wetlands—wetland banking. Note that this market-based approach still sits upon some regulatory framework, in this case to place a cap on wetland loss.

Now let's think about some criteria upon which to judge these approaches: cost-effectiveness, fairness, the ability to incentivize innovation, and the ecological outcomes. The No Net Loss approach gets the check mark for cost-effectiveness, because some wetlands are cheaper to protect and others carry a high opportunity cost of protection. Imagine a wetland that is biologically poor, and functionally irrelevant, but that sits in the prime spot for a children's museum that will provide a huge social benefit. Farther away there is an area that was a wetland a century ago but was drained for agriculture. This area can be restored very cheaply and will provide important functional and biological services if restored. Under the strict no-conversion regulation, society would lose out on a huge social benefit (children's museum *and* a new functioning wetland). Under the No Net Loss institution, society gets huge benefits for little cost. Check goes to the market-based approach.[6]

What about fairness? Let's think about the NEBNSGE Co. again. Is it fair if the company is able to buy its way out of conserving a given wetland, when a small holder farmer cannot do so? Under strict regulation everyone is treated equally. Let's give the check to the regulatory approach. Innovation is similarly easy.

---

[6] Yes, we stacked the deck here; this scenario is highly improbable but not impossible.

Under strict regulation there is reason to innovate on the buyer's side, thereby avoiding the need for conversion in the first place. But that is also true under the market-based policy. Additionally, under the No Net Loss rule, with credits and banking there are reasons why sellers would try to restore better wetlands—incentives exist to make wetlands more attractive or less costly to buyers and banks. Check goes to the market-based approach.

Finally, what about the ecological outcomes—the biodiversity, the uniqueness, and the functioning of wetlands? Well, of course we can imagine a scenario as described earlier where a restored or constructed wetland is better than the one it replaces, but the evidence to date generally suggests otherwise. We give the check to the strict regulation and suggest that ecological outcomes would, on average, be better by protecting existing wetlands, full stop.

We could judge these institutions against other criteria, but you'd still see that under some considerations one would be better than the other. Market-based instruments are becoming popular approaches in conservation due mainly to their cost-effectiveness, ability to incentivize innovation, and in some cases political feasibility (as compared to instituting a new law). But some of the necessary requirements for market approaches make them difficult to implement. For example, remember back in Chapter 4 where a market good needs to have "excludable" qualities? That implies ownership. Trying to coerce common-pool resources or public goods into market goods often requires the establishment of property rights. As we've discussed before, not an easy thing to do. See **Box 6.1** on p. 130 for a few other obstacles to the use of market-based approaches in conservation.

We hinted earlier that the market-regulation dichotomy is a false one for categorizing institutions. We already saw that it is false because market interventions almost always require some regulatory structure (e.g., granting rights, upholding policies, supporting markets, enforcing laws). However, it is also false because across the world, humankind has sometimes organized itself to protect, conserve, and supply common-pool resources and public goods without any formal market or regulatory institution. The pioneering work of Elinor Ostrom (1990) showed what enabling conditions were for such systems of self-governance to be successful; but perhaps it is enough to say for now that currently there are more than 450,000 collective resource management groups around the world governing communal resources without formal markets or regulatory support (Pretty 2003). These groups are collectively managing their watersheds, irrigation systems, forests, fisheries, and so on.

So we saw how different institutions are set up, the diverse forms they can take, and where and under what conditions some are better than others. We also saw how, although in essence they are set up to correct market and information failures, they are not always successful and sometimes cause as many problems as they aim to overcome (this would be *institutional failure*). But let's spend a minute to dig a little bit deeper into one of the most popular institutions at the moment for internalizing the negative externalities of our decisions: *payments for ecosystem services*.

> **Payments for ecosystem services (PES)** are institutional arrangements in which a willing buyer compensates a willing seller for the verifiable delivery of an ecosystem service.

BOX 6.1 **Some Complications with Market-Based Institutions for Conservation**

Along with the real gains in conservation that market-based interventions can bring, there is a lot of rhetoric, greenwashing, and hocus-pocus. Below are a few of the core obstacles that market-based interventions have to overcome.

■ **Property rights:** The establishment of property or use rights over a good or service is usually critical. Sometimes this isn't easy. Think about the services provided by a wetland that changes in size by a factor of 10 between the wet and dry seasons; in other words, different people might own the wetland depending on the season. If property rights are not clear, then incentives to invest in sustainable uses are likely to be weak, because buyers can't be sure of who the sellers are and whether they will continue to provide a service into the future.

■ **Measuring and monitoring:** It is difficult to measure and monitor many ecosystem services. Is this the patch of forest that is providing water regulation benefits, or is it the neighboring one? Is this section of reef providing stable fish stocks, or do they actually come from elsewhere? Getting the measurements and monitoring right is very important for buyers of services to know that they indeed get what they pay for.

- **Correct pricing:** For price-based mechanisms, finding the correct price level to incentivize conservation is not likely to be straightforward. Offer too little and the supply will be too low. Offer too much and the mechanism becomes cost-ineffective. For example, in Costa Rica, the land under forest conservation contracts is more likely to be low-value agricultural land on steep slopes and inaccessible, suggesting that the buyer (the government) is overpaying.

- **Cultural hurdles:** Price and quantity-based mechanisms assume that a market institutional setup is common. In some places, assigning property rights or paying people to behave in a certain way may not be culturally acceptable. For example, in some cultures the idea of "owning" land or having exclusive rights might not exist.

Of course, there are more obstacles than these. Some of them we described in this chapter. There are also many impediments to making regulatory approaches effective and legitimate, and the world of environmental policy deals daily with those.

Payments for Ecosystem Services in Arden and the Real World

Let's return to Arden. We mentioned in Chapter 5 that Isabella derives satisfaction from the stream that runs through her land in a variety of ways. But unlike her trees, the stream also passes through land that belongs to her neighbors. So Isabella's management of her section of the river has consequences for others. For example, remember that Xavi has already stated the value he places both on general aesthetics (Chapter 2) and on bird habitat (Chapter 5) provided by the stream. Unfortunately, if Isabella clear-cuts her forest, the increased runoff from the newly cleared areas will result in siltation of the river, which will reduce water quality for fish, birds, and Xavi's daily swim. Such a move could also increase flooding on Xavi's land (remember the goals of China's Sloping Land Conversion Program?).

What can Xavi do about this situation? He could of course talk to Isabella and ask her not to chop down her forest. However congenial Isabella may be, Xavi's pleas may not be enough for her to change her decision in the face of the profit she would make by selling the timber. Xavi could also appeal to the government, either to enforce any existing laws regarding forest clearance or to lobby for changing relevant laws in a way that is favorable to him (and to conservation).

There is another option: Xavi may simply decide to pay Isabella not to cut down her forest. Such an approach would be a payment for ecosystem services (PES). We examined this sort of institution in Chapter 4, showing how a carbon payment might affect Isabella's cost-benefit analysis. But now that we're

in the Institutions chapter, let's take a minute to more formally define a PES.

Sven Wunder (2007) at the Center for International Forestry Research (CIFOR) has articulated five conditions necessary for a PES:

1. There is at least one buyer of a service.
2. There is at least one seller of a service.
3. The buyer pays the seller for a well-defined ecosystem good, service, or agreed-upon proxy.
4. The payment is made conditional upon the verifiable provision of the service being bought.
5. Both the purchase and the sale of the service are voluntary.

Now this is a long list, and these are ideal conditions for a PES. Condition 4 is particularly difficult in some cases given the time lags required to see changes in the flows of ecosystem goods and services. Condition 5 is often modulated by government involvement and decisions that take place above the individual. Nonetheless, these are good guideposts for understanding PES.

In our Arden case, Xavi (the buyer) is up for voluntarily paying Isabella (the seller) to provide . . . well, what? It is often difficult to both clearly define an ecosystem service and be able to effectively monitor that it is in fact being delivered. Xavi knows that he values the river's water at its current level of clarity and cleanliness for the benefits it provides. So he is willing to pay Isabella to refrain from worsening the river's current condition. He might use tools (chemical testing, Secchi disk)

to monitor the stream's water quality to ensure that it is in fact staying clean and that he is getting what he is paying for. Alternatively, he may just assume that as long as Isabella maintains her forest in its current undisturbed condition, the stream's water quality will remain unchanged. This kind of simplifying assumption about land cover and ecosystem service provision is fairly common, given how complicated and expensive it is to measure the actual ecosystem service in question (again the problem with condition 4 in the PES list). This is especially true when the buyer is a national government dealing with a diverse group of multiple sellers over a wide, difficult-to-monitor geographic area.

How are prices for ecosystem services determined in PES schemes? Xavi understands how much value he places on the benefits delivered by the stream, and Isabella knows what she stands to gain from harvesting the timber on her land. Neither individual knows the other's figure, however. This is information asymmetry (remember from our earlier discussion?). In the case of Isabella and Xavi, they can haggle using their own WTP/WTA and likely arrive at a price that both consider acceptable. However, many PES schemes consist of a large number of landowners selling services (or land cover) to one buyer, typically a government agency. In such cases, how does the government know how much to pay each individual? Schemes that pay everyone the same amount overlook the big differences in opportunity costs that landholders have for conserving their land. For example, it has been shown that the Costa Rican government is often overpaying for ecosystem services. A large number of applicants to their national-level PES program (in which the government pays farmers to conserve forest) are farmers whose

forest land is of little use for agriculture. For these farmers, the opportunity cost is low; the government could have paid them less for the same outcome (Ferraro 2008). On the other hand, more precisely targeting the sellers who have the lowest opportunity costs—by using an auction, for example, where farmers bid for a price they'd be willing to accept—is itself a costly thing to do.

OK, we hope you see that conservationists rely on a range of institutions to help overcome some of the market and information failures we've been discussing throughout the book. Some of these institutions seem to be working well for ecological outcomes (Costa Rica's PES), some seem to be working from a cost-effectiveness standpoint (No Net Loss), and the jury is still out on others (like product certification schemes; see Blackman and Rivera 2011). Some institutions seem to be effective from both a social and ecological standpoint (Namibia's Community-Based Natural Resource Management Program; see **Box 6.2** on p. 136), and some have so far been a disappointment (the UN's REDD program, which has not yet delivered a real global market for forest carbon).

So things are far from perfect, and in some cases we can add institutional failure to market and information failures. After decades of rigorous examination of institutions governing public and common-pool resources, Elinor Ostrom concluded that there is "no panacea," no silver bullet, in this sphere. We must continue learning and experimenting with institutions to better understand what conditions lead to success. This is especially true for institutions that operate at broad scales with many actors; these more realistic situations are what we tackle in the next chapter.

BOX 6.2 Namibia and the International Transfer of Conservation Benefits

Namibia's Community-Based Natural Resource Management (CBNRM) program is an institutional arrangement that governs the sustainable management and use of wildlife and other natural resources on 160,000 square kilometers of communal lands. The program began in 1998 with the establishment of the first communal conservancies, and it currently involves about 234,000 rural inhabitants who benefit from natural resources on their communal lands. Prior to the mid-1990s, these natural resources had been official property of the government—a legacy of apartheid-era policies—but an institutional change in the form of a new law transformed the situation. Communities benefit from wildlife and plants in a variety of ways: by revenues and employment from joint ventures involving tourism and trophy hunting, by their own hunting of wildlife for meat, and by the sale of plants for use in a variety of products. Some of these benefits, such as the hunting of wildlife for meat, are locally generated and consumed, but many of the others involve international beneficiaries who pay Namibians (in valuable foreign currency) in exchange for the benefits their biodiversity produces. Robin Naidoo and his coauthors (2011) have studied this program as a PES scheme in disguise, and you can spot the mix of market-based and regulatory approaches that we describe in the chapter.

Let's unpack these benefits a little further. In terms of nature-based tourism, local communities typically enter into joint-venture agreements with private companies to build and manage lodges and other infrastructure to support international tourists. In return, communities negotiate deals with the tourism operators that pay them a flat percentage of total profits and also specify that local community members must be hired as staff when possible. So this is a one-buyer, one-seller agreement with an agreed-on price for agreed-on ecosystem services. Hunting operations function in much the same way, although typically there are fewer employment opportunities and revenues are lower than for tourism. However, meat from the hunted trophy animals (depending on species) must be turned over to the conservancy, and that is viewed as a key benefit.

Some of the most intriguing transactions in the CBNRM program involve deals negotiated for products made from plants. For example, the devil's claw plant (*Harpagophytum zeyheri*) produces an analgesic compound used in painkillers and anti-inflammatories, and a number of conservancies where the plant prospers have agreements with pharmaceutical companies to deliver plant parts to them for a negotiated price. Similarly, resin from several species of *Commiphora,* camphor plants found in northwestern Namibia, is used in perfumes marketed in Western countries; the first 5-ton harvest, worth USD $50,000, was sold to the French cosmetics company BeHave in 2007 by Marienfluss Conservancy.

(continued)

BOX 6.2 continued

This is a rather extraordinary series of markets that bring wealthy and sophisticated Western tourists and businesses together with some of the most marginalized and impoverished people on the planet. So it is a market-based approach but one that, crucially, relies on a change in institutional structure to allow it to flourish. In so doing, it has not only resulted in significant increases in wildlife populations—which in the 1980s had been poached to historically low levels—but has also generated several million dollars for conservancies (see the figure below). In today's increasingly globalized world, Namibia's CBNRM program offers a positive example of how a combination of progressive institutions and international markets can result in tangible conservation and livelihood benefits.

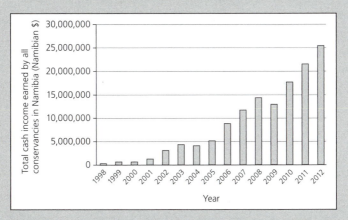

Total cash income earned by all conservancies in Namibia, 1998–2012. Values are in Namibian dollars (in May 2014, 1 US dollar was worth approximately 10 Namibian dollars).

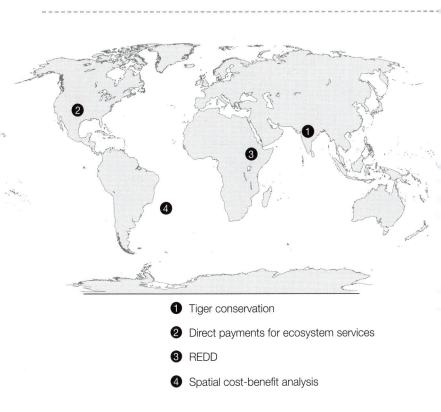

1 Tiger conservation

2 Direct payments for ecosystem services

3 REDD

4 Spatial cost-benefit analysis

Scaling Up and Getting Real

Managing Complex Landscapes with Multiple Goals

So far in this book we've been examining the economic trade-offs among competing land-use options in Arden. Here's a secret: Arden's not real. It includes only a handful of properties, a few interestingly named people, and some very simple choices. To be truly useful for conservation, economics must inform the real, complicated, broader-scale world. This world involves landowners like Xavi, but also many others, across large areas. It also involves regional governments, watershed authorities, and businesses. These actors exert economic and social forces on conservation from next door and from halfway around the world.

In this chapter we try to scale up and get a bit more real. In particular, we examine how economics can help us make efficient use of scarce resources when addressing "large-scale" activities (i.e., beyond the unit of an individual property): designing networks of protected areas, optimizing regional economic returns from ecosystem services, and managing landscapes for competing conservation and non-conservation goals. We will

reach back into earlier chapters to weave various threads together in a final version of Arden. For each, we'll try to show how economic concepts clarify conservation issues and how they point toward solutions. To help pull it all together, we'll imagine three scenarios of future development in Arden and estimate the economic costs and benefits of each. These scenarios will illustrate how economic information can be, and increasingly is, used in everyday conservation planning decisions.

Arden's Apex Predator

But first, let's add one more wrinkle to Arden. Remember that Arden's landscape already produces a range of benefits to Xavi and his neighbors, and that these can be in conflict with one another (e.g., the benefits that Xavi derives from bushmeat hunting are incompatible with the benefits Isabella will derive by clear-cutting the forest on her land). In all this, we neglected to mention that Arden is also home to the world's largest cat: the tiger. *Panthera tigris* in Arden is an obligate forest species (fancy conservation term for "this thing needs forests"); it ventures into agricultural areas only during dispersal periods and for the occasional nocturnal cow snack. Because tigers live there, Arden is an area of global conservation concern (and probably the only place where tigers and people named Xavi and Isabella coexist).

Back in the real world, tigers' range and numbers have decreased by about 90% since the beginning of the 1900s. There are now more captive tigers in the United States than there are

left in the wild.[1] As a wide-ranging top predator, they require large areas of habitat and prey, and so in landscapes where human settlement is increasing, tigers frequently come into conflict with people, resulting in deaths to livestock or to humans.

Threats to Tigers and Other Biodiversity in Arden

Given this additional wrinkle, what are some of the major conservation issues in Arden? Let's first consider threats to the persistence of tigers and other biodiversity in the Arden landscape. The single greatest threat to global biodiversity is land-use change: the conversion of natural ecosystems like forests to human systems like farms or cities. Arden's forests are home not only to tigers but also to the myriad other species that ultra-diverse tropical forests house. As mentioned, tigers are wide ranging and require large areas of forest in order to survive. By successfully conserving "umbrella" species such as tigers, other species that require smaller patches of forest should be conserved along with them. In the case of Arden, agriculture is the dominant human land use, and to meet demand for food from a growing human population, both inside and outside of Arden, new agricultural land must be carved out of the forest. We've zoomed out our map of Arden to show the current

[1] According to the World Wildlife Fund in 2012, there are roughly 5,000 tigers in captivity in the United States alone, compared to a bit more than 3,000 in the wild worldwide. Of those captive cats, more than 90% are in backyards, urban apartments, or private reserves (not accredited zoos). Whoa.

resulting pattern of land use (**Figure 7.1**).Unless this conversion of forests to agriculture is curbed, tigers and many other forest species will have a dim future in Arden.

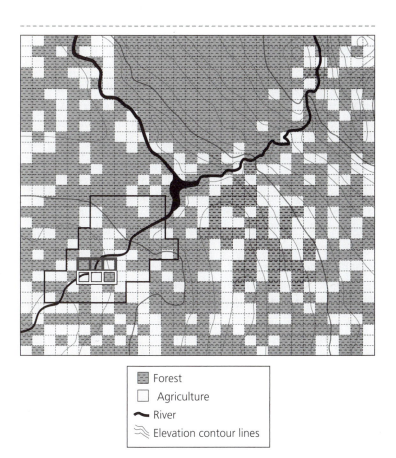

Forest
Agriculture
River
Elevation contour lines

FIGURE 7.1 Land-cover map of Arden. Isabella's and Xavi's properties are marked for reference, as in the outline of Figure 5.1.

Another major threat to the current distribution of species is global climate change. As carbon dioxide emissions continue to rise, rapid changes in temperature and precipitation patterns will disrupt the environments that species have adapted to for millennia. At least some species will be unable to adapt or to relocate quickly enough, and will therefore perish. A new international institution, REDD (Reducing Emissions from Deforestation and Degradation), has been touted as a way of both mitigating against increasing carbon dioxide emissions and conserving tropical forest biodiversity. REDD works by rewarding countries that reduce deforestation rates from baseline levels, thus providing incentives to maintain standing forests and the biodiversity they harbor (Gardner et al. 2012).

A final and vexing threat to tigers is the traditional use of their body parts in medicine and ornamentation, which goes back thousands of years in Asia. In early times the number of tigers killed for these products was low relative to their global population, but as humans and their technological capabilities increased, pressure on the nine geographically distinct subspecies of tigers began to take its toll. The Bali subspecies was driven to extinction in the 1930s, and the Caspian and Javan subspecies disappeared in the 1970s. The South China subspecies is also likely extinct in the wild, and remaining tiger populations everywhere are under severe pressure.

When a conservation organization or government tries to conserve a species like the tiger in a place like Arden, these are the types of issues that must be considered. Many of the drivers, threats, and opportunities associated with biodiversity conservation arise from markets and human actions in faraway places. In addition, different organizations will have different goals for how to manage landscapes. Our goal in this chapter is to

illustrate how the economic concepts we've covered so far can help conservationists understand the forces stacked for and against conservation.

The Demand for Tiger Body Parts

Let's first consider how demand for tigers creates forces that affect their conservation. There is a sense that increasing levels of wealth in China, in particular, have spurred demand for tiger parts and thereby led to increased poaching to supply this (illegal) market. However, actual evidence for this is pretty sketchy, and the relative effects of wealth, demography, culture, and potential substitutes on demand for tiger products aren't well understood. Because consumption of (wild) tiger products is illegal in China, collecting information on the tiger-parts trade is naturally very difficult, and the few attempts have therefore involved elaborate undercover investigations (Environmental Investigative Agency 2013).

A choice experiment, which we discussed in Chapter 5, might be a more straightforward way to understand what drives demand for tiger products. Nick Hanley and his friends at the University of Stirling and in Tanzania have used choice experiments to figure out what influences illegal hunting of bushmeat in the Serengeti ecosystem (Moro et al. 2013). Using this method, they avoided having to ask sensitive questions on actual levels of illegal hunting, since scenarios are hypothetical. The authors found that alternative income sources and better enforcement both reduce illegal hunting, and that wealthier households were less likely to hunt and also less concerned about enforcement than poorer households. These results

show that effective policies to deter hunting should be targeted differentially across the region, allowing more pinpointed (but probably more costly) interventions.

A similar approach for tigers would need to survey a random sample of people in areas where tiger parts are used in traditional medicine and as ornaments. Using contingent valuation or a choice experiment, we could estimate the relative effects of wealth, culture, and other variables on demand for tiger parts as well as the potential for substitute products (from nonthreatened animals, or synthetic) to satisfy some of this demand. As with the Tanzania study, these results would allow the development of policies based on actual supply and demand market data, increasing the likelihood of their effectiveness. Since this is a book on economics, we should also be clear that running these types of surveys themselves costs money. We presume that the benefits of the good data obtained from such an approach justify the costs . . . but sometimes, as old-time crooner Bob Seger put it (in a different context, for sure): we ain't got no money. Chapter 5 discussed different ways to collect economic data that are less costly than field surveys.

Back to the tiger example: Collecting these supply and demand data could help in the efficient allocation of conservation resources. To illustrate, recall an important concept that we first introduced in Chapter 3: income elasticity of demand. To recap, most goods are elastic; that is, demand for them generally increases as a person's income increases (or similarly as the price of the good decreases). However, the devil is in the details. Increases in demand vary greatly across all goods—and may do so at diminishing rates. In the case of the illegal market for tiger products, this is a big deal. Suppose, for example, that our choice experiments show a strong and increasing relationship

between demand for tiger body parts and income, and that this effect is stronger than cultural or other demographic characteristics. Anti-poaching efforts would therefore need to track the overall economy, stepping up when times are good, jobs are strong, and incomes are increasing. On the other hand, if demand turns out to be largely inelastic (i.e., independent of changes in income) and is dampened a lot if credible substitutes are available, then making alternative products available and attractive should be a key focus for conservationists. Whatever the specific outcome, conservation strategies would be informed and strengthened by gathering some economic data on the market for tiger products.

The Demand for Living Tigers

Benefits from tigers are not restricted to body parts from dead animals. There is also a demand for living tigers in the wild, largely from faraway Westerners who want to know the species will survive or who are willing to pay to travel overseas for the chance of seeing a tiger in its natural habitat. Both values can provide real money for tiger conservation, depending on how these benefits are captured (which is a job for institutions; see Chapter 6).

Ecotourism is a class of tourism that, according to the International Ecotourism Society, involves "responsible travel to natural areas that conserves the environment and improves the well-being of local people."

In the case of tourism, the government and private operators have an economic incentive to conserve tigers (a major tourism draw in certain reserves in India and Nepal), because these entities are the most likely to commandeer the majority of cash benefits from

tourism. There are many *ecotourism* schemes that intend to divert at least a portion of the money to local landowners, villagers, and other stakeholders who typically bear the highest cost of providing tourism opportunities (i.e., the opportunity costs of conservation; see the next section). Yet these efforts have yielded mixed results at best, with locals often seeing very little of the actual benefits from tourism, certainly not enough to offset their opportunity costs.

Consumers who care about tigers may also be willing to pay more for products (timber, palm oil, etc.) that are certified as having been produced through "tiger-friendly" means. Brendan and Robin's paper on tiger-friendly margarine is one example of quantifying consumers' additional WTP (Bateman et al. 2010). Their research suggests that consumers in the United Kingdom were willing to pay between 15% and 56% more for such a product.

These "indirect" ways of contributing to conservation (i.e., through conservation-friendly industries) have been criticized as inefficient (Ferraro and Kiss 2002). The argument against them, essentially, is why not pay directly for what you want (tigers in the wild) rather than for an affiliated product or recreational opportunity that may or may not be very strongly linked with the conservation objective? Of course, people who care about biodiversity and wild nature have been making donations to conservation groups for decades. More recently, emerging PES-type schemes that link *direct payments* to biodiversity have been gathering steam. As we saw in Chapter 6, these PES

> **Direct payments** are payment schemes that pay directly for a conservation outcome of interest (biodiversity, carbon, water, etc.). Payments for ecosystem services are an example of direct payments.

schemes link the users of an ecosystem service with those who provide it, creating an economic incentive for producers to continue their conservation activities. One interesting example comes from Kenya, where a private company (Wildlife Works) has directed money from investors interested in conservation of wildlife to conserve a critical corridor that connects Tsavo East with Tsavo West National Park. A key difference with this scheme is that investors pay for particular biodiversity outcomes; payments are contingent on these outcomes being independently verified via on-the-ground monitoring. This is one of several examples of "Wildlife Premium Markets" that operate over large spatial scales (Dinerstein et al. 2012).

The Costs of Tiger Conservation

A variety of benefits are derived from wild tigers, both living and dead. We now turn in more detail to the costs of conserving wild tigers.

The opportunity costs of a particular land use can be estimated by the forgone benefits associated with the next best land use. In the case of Arden, we know from Chapter 2 that on Isabella's property, converting forests to agriculture results in a net profit of $400 (or $200 per parcel), so as a first-cut estimate we might use this value for the opportunity cost of conservation on all forest parcels in the landscape. However, and as with biodiversity, we know that the distribution of opportunity costs (i.e., agricultural profitability) is spatially variable. Certain parts of the landscape have better soils, resulting in taller forests with more timber and better potential for crop cultivation once they have been cleared. Other parts of the landscape are rugged and

mountainous, which make them more difficult for both timber harvesting and agriculture. Still others are suitable from a biophysical point of view, but they are far away from existing roads; therefore, markets for agricultural products are highly inaccessible.

If we are to plan for conservation at the landscape scale, we need to understand how properties vary in these opportunity costs. **Figure 7.2** shows how productivity of coffee—which

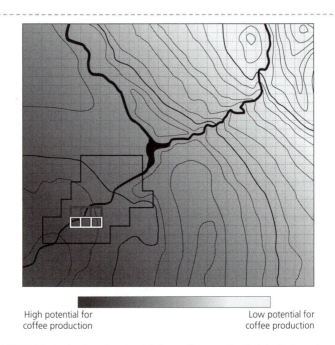

High potential for
coffee production

Low potential for
coffee production

FIGURE 7.2 Agricultural potential for coffee production in Arden; the darker the shading, the greater the potential. Soils and climate become increasingly suitable as elevation decreases from the northeast corner toward the southwest corner of the map.

is directly related to potential agricultural revenue and therefore the opportunity costs of conservation—is distributed geographically across Arden. We can see that soils and climate improve for coffee production as altitudes get lower, toward the southwest corner of our map.

How could we estimate the opportunity costs we describe above? One method is to use a hedonic pricing approach (Chapter 5) that builds a statistical model using the factors like soils, topography, and roads to predict known land values. Another approach integrates the probability of a parcel of land being deforested with the expected profitability of agricultural and/or livestock production on these converted lands. These methods produce spatial "surfaces" of conservation opportunity costs (Adams, Pressey, and Naidoo 2010; Ban and Klein 2009).

If land values are unknown, incomplete, or of uncertain accuracy, the use of auction-type mechanisms to reveal the opportunity costs of conservation may be appropriate. We touched on auctions briefly in chapters 3 and 6. The logic behind the use of auctions is that individual landowners are the ones who best understand their opportunity costs of conservation. However, landowners typically have no incentive to reveal their true willingness-to-accept values to potential buyers, and they have every incentive to inflate those values. Think about it: if you knew there was someone very keen to buy up all the bikes in your neighborhood who had a huge wad of cash to do so, you might consider selling your bike, but you'd probably ask for far more than it's worth to you. Ditto for conservation organizations trying to make land purchases or negotiate conservation easements (Chapter 6). Auctions stimulate competition among landowners to receive potential payouts; auctions therefore provide landowners with an incentive to reveal their true opportunity costs.

Another major class of conservation costs relates to damages from wildlife. People living near tigers are well aware of the dangers: tigers kill livestock and occasionally people. The fear factor of living near these intimidating cats is an equally important consideration. Although fear of tiger attack is difficult to assess, the monetary costs of tiger predation can be more readily estimated from predation events and the lost value of a cow. In addition, wild animals that tigers prey on may be dangerous themselves (buffalo, gaur), and they may also damage crops or property of nearby residents. Damage costs tend to be highly unevenly distributed across time and space; residents may have lived in a tiger habitat for years without ever seeing any sign of one, then suddenly may have several goats or cattle killed within a span of a few weeks. All of these costs must be borne by locals living near conservation landscapes that sustain tigers and other species that may be a nuisance—or an outright danger—to people.

There is a huge and interesting literature on strategies to reduce these costs and/or to make them more palatable to those who are affected (Agarwala et al. 2010; Dickman, Macdonald, and Macdonald 2011). For example, a study on wolf conflict in Wisconsin found that, as expected, people who had lost domestic animals to predation were less tolerant of wolves than those who hadn't experienced losses. Interestingly, however, people who had received compensation were actually more likely, rather than less, to support wolf population reductions than those who had not received any compensation. Other factors, such as attitudes and the socioeconomic profile of respondents, appeared more important than financial considerations, indicating the complexity of dealing with emotive issues such as human-carnivore conflicts (Naughton-Treves, Grossberg, and Treves 2003).

Systematically Evaluating the Costs and Benefits of Tiger Conservation

All right, we now know a little bit about the various costs and benefits of having tigers in Arden. Still, how can this information help government agencies and others in managing this area for particular goals? We'll illustrate by evaluating the outcomes of three particular types of management: thinking only about agricultural production; thinking only about tiger conservation; and trying to balance the economic costs, benefits, and trade-offs associated with both.

We know the pattern of potential revenue from coffee production in Arden (Figure 7.2), but what about tigers? Where can they be found? Conservation planners typically assign some biological value to all spatial units in a landscape, based on where tigers have been seen, expert opinion on the suitability of particular land-use types for tigers, or scores from fancy habitat suitability models. In this case, let's say that biologists have done a good job collecting tiger observations from camera traps,[2] hair traps, and track locations. Modeling this occupancy across the landscape, based on vegetation cover, prey locations, and disturbance, suggests that the suitability of tiger habitat is distributed as shown in **Figure 7.3**.

So let's now start with a situation in which an agency is concerned only with maximizing agricultural production in Arden. This is still the de facto management plan of many rural areas on the planet. Landowners must weigh the potential

[2] If you don't know what these are (or even if you do), search the Web for "BBC Wildlife Camera-Trap Photo of the Year." Brendan has his own camera trap for personal use in the woods behind his house, but he has never won.

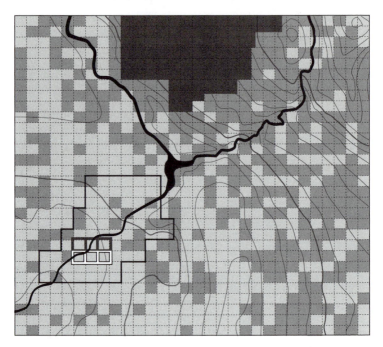

FIGURE 7.3 Tiger habitat suitability, with darker cells indicative of habitat with higher suitability. The core habitat area in the north is the most suitable; all other forests are moderately suitable, and agriculture is least suitable.

benefits of agricultural production (a function of crop prices and the biophysical suitability of their land) versus input (labor, fertilizer) and transportation costs (effort required to get product to market). When benefits exceed the costs, rational landowners will continue to farm existing croplands and will convert forests to farmland. **Figure 7.4** shows what Arden might look like under such a scenario. A lot of forest

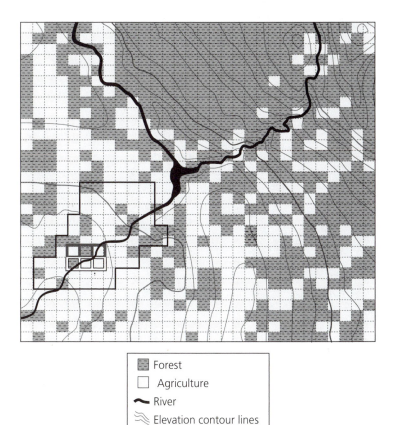

Forest
Agriculture
River
Elevation contour lines

FIGURE 7.4 Arden landscape under expected agricultural expansion. This is probably the "business as usual" scenario. Many forest cells have been converted to coffee, especially in the more productive southwest portion of the landscape.

has been converted to coffee, mainly in the southwest, where coffee productivity is highest. Much agriculture remains even in lower-productivity areas, although agricultural areas where costs exceed benefits (e.g., due to soil nutrient depletion) have been abandoned and have reverted to forest.

How might maximizing agricultural production affect the things we care about in Arden? **Table 7.1** tracks all this. We can see that the agricultural benefits from this landscape total $232,000 per year—that's the net revenue from coffee production of all farmers in Arden. The biologists' models predict the landscape will support 200 tigers. And summing up simple scores across the landscape, we calculate an aesthetic score (to capture the less tangible benefits that Xavi and Isabella derive from being in the forest) of 508. Not bad.

TABLE 7.1 Tabulating monetary benefits, tiger numbers, and aesthetic values for three landscape-management scenarios in Arden.

| | SCENARIO | | |
	Agriculture	Tigers	REDD and tourism
AGRICULTURE PRODUCTION VALUE	$232,000	$144,000	$110,000
REDD PAYMENTS	-	-	$170,000
TOURISM REVENUE	-	-	$110,000
TOTAL	$232,000	$144,000	$390,000
TIGER POPULATION (number of tigers)	200	220	240
AESTHETIC SCORE (points)	508	614	680

Here's something important, though: while the spatial pattern of inherently good and bad land for farming changes slowly (Figure 7.2), at least on human time scales, the prices of inputs (like fertilizer and labor) and the price of coffee itself can change quickly. Just as important, those changes are due to global market forces that are beyond the control of Arden-dwellers. Coffee prices in the real world have been notoriously volatile, whipsawing farmers into different land-use decisions and causing remarkable changes to landscapes. Similarly, in Brazil the prices of agricultural commodities such as soybeans and beef are as good or better predictors of deforestation than factors such as government policies or conservation efforts (Hargrave and Kis-Katos 2013). The same is likely to be true for commodities such as palm oil: as demand for, and therefore the price of, palm oil has increased, deforestation in Southeast Asia has increased as forests have become replaced by oil palm plantations (Koh and Wilcove 2007). The same economic forces can sometimes lead to positive conservation outcomes. The great hardwood forests of eastern North America, which were largely cleared by the end of the eighteenth century, have rebounded greatly since then as agricultural production became cheaper in other parts of North America and the world.[3] In both positive and negative instances, economic forces outside of the spatial area in question can and will play a dominant role in shaping land-use patterns.

Moving on, let's now consider the improbable case where the government of Arden decides that its dominant policy

[3] Apart from the conservation benefits, the forest recovery has made Brendan and Taylor very happy; 75% of their home state of Vermont is forested, making it a pleasant place to live and play (and a major reason why it has taken us so long to write this book).

objective is to maintain and enhance tiger populations. This policy protects all remaining forested areas from timber harvesting and agricultural expansion, and it includes a law that requires landowners who own more than 100 hectares of land to maintain 50% of their property as forest. In the real world, countries such as Brazil and Paraguay have laws—though often weakly enforced—that stipulate such levels of forest protection.

How does this scenario compare to the "maximize agricultural production" scenario described previously? From our zoomed-out perspective, the landscape will look very similar to the current case in Figure 7.1, since major forest clearing would stop (and some landowners would be compelled to restore forest to achieve the 50% threshold). Because this scenario retains more forest than the agricultural one does (Figure 7.4), the tiger population would be 10% higher across Arden (Table 7.1). Increases in forest cover also result in Arden's becoming a bit more aesthetically pleasing. However, this change comes at a cost: a lot less agricultural production ($144,000).

The opportunity cost of increasing tiger populations by 20 in this scenario is the $88,000 in lost agricultural production, although there are probably additional damage costs due to tigers snacking more often on cattle. Incurring these costs would require strong political will on the part of Arden's government and citizens. Even if Arden is willing to absorb these opportunity costs, it will not assure tiger persistence on the land. There are numerous examples around the world of the so-called empty forest syndrome, where intact natural habitats remain but are devoid of large wildlife species due to poaching of animals for meat, skins, and sport. In a highly publicized case in 2007, India resurveyed the jungle haunts of its largely intact system of tiger reserves and found that the estimated number of tigers living

within it was 60% lower than the number estimated in 2002. In a number of reserves there did not appear to be any tigers remaining. Despite tigers' protected status, the economic incentives for poachers to enter reserves and kill tigers for export to the Chinese black market was very high. For example, the price of a tiger skin has been estimated to fetch as much as USD $35,000, even a fraction of which is a huge windfall to an impoverished local villager. Therefore, habitat protection alone is no guarantor of the persistence of species that are at risk from economic forces.

An approach more likely to lead to long-term, sustainable conservation of tigers on the landscape is one that evaluates and realizes their tangible financial values alongside those of other competing economic interests. In our third scenario, let's examine what happens to both tiger conservation and economic gains to landowners when a number of ecotourism enterprises centered on tigers have sprung up on the landscape, and where a *REDD*-like system rewards the conservation of forests. Given current agricultural prices and carbon payments of $5 per ton of carbon, Arden residents would choose to conserve an additional 172 units of forest (**Figure 7.5**; Table 7.1), compared to the agricultural scenario (Figure 7.4). We would also predict an increase in tiger populations of 40 individuals. The lost agricultural land means that coffee revenue drops to $110,000 landscape-wide, but REDD payments more than make up for that. In fact, this scenario produces the highest economic

REDD stands for Reducing Emissions from Deforestation and Degradation. These United Nations–agreed policies are intended to reduce carbon emissions through the conservation and management of forests in developing countries.

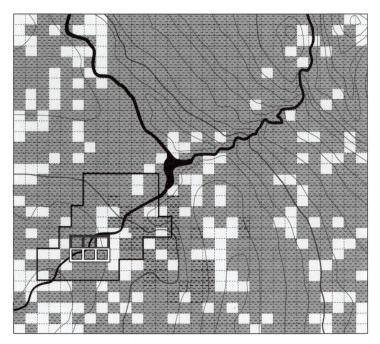

Forest
Agriculture
River
Elevation contour lines

FIGURE 7.5 Arden landscape under scenario of REDD programs and ecotourism. Given these incentives to conserve and restore forest, many formerly agricultural cells have been restored to forest. For the landscape as a whole, the forgone agricultural income is compensated for by REDD and tourism revenue.

value for the landscape while increasing tiger numbers and having the highest aesthetic score (Table. 7.1). Pretty good!

Despite its high economic value, there would still be issues with implementing this vision in Arden. One problem is that REDD payments are made to landowners regardless of the value of their forests to tigers, so we can't be sure that carbon payments conserve tigers. That has led many to argue that biodiversity considerations and REDD need to be more tightly linked (Strassburg et al. 2010, Gardner et al. 2012). Luckily, others have shown that tiger habitat tends to be more carbon-dense than non-habitat forests within tiger ranges (Wikramanayake et al. 2011). Consider also the development of ecotourism enterprises specifically marketed toward tiger excursions. These include companies that operate in the core protected area as well as two run by private landowners whose large properties provide opportunities for regular tiger sightings. We won't go into the economic decisions involved in opening an ecotourism lodge; others have (Kirby et al. 2010). The point here is simply that a well-run, profitable enterprise will provide additional economic incentives for the landowners to conserve forests and tigers. Landowners would need to make a profit in order for ecotourism to outcompete the conversion of their forested land to agriculture. In terms of smaller landowners living around the core protected area, "spillover" sightings from tigers may be reliable enough to develop ecotourism ventures, but it is more likely that agreements between landowners and ecotourism companies to conserve tigers, based upon verified evidence of tiger occupancy, may be in the economic interest of both parties. Nevertheless, the difficulty of ensuring that some of the profits from ecotourism make their way into local hands is widespread, and there are many examples

of nature-based tourism where a theoretical incentive for conservation exists, but in practice, profits remain in the hands of tourism operators rather than those whose decisions actually shape conservation on the ground (West, Igoe, and Brockington 2006).

This chapter has shown how a landscape-level, rather than property-level, consideration of costs and benefits of conservation can inform broad-scale management and planning. In addition, we've highlighted the critical role of institutions to capture values and make them real in the cost-benefit calculus of landowners and others. These institutions help address the fact that in many places, the people who benefit the most from conservation are those who live far away; in addition to the examples we've already raised, other studies have shown that people in rich, developed countries are willing to pay to conserve species or ecosystems in developing countries that they may not ever visit themselves (Horton et al. 2003; Morse-Jones et al. 2012). Capturing these benefits can offset the opportunity costs of local landowners and communities, who must bear the brunt of conserving forests and tigers.

As we understand more about the benefits that natural ecosystems contribute to human welfare, the development of additional institutions that capture these values will likely further tilt the balance from exploitation to conservation. And while Arden doesn't exist anywhere on our planet, the issues that we've discussed are played out on a daily basis in many different parts of the world. Understanding these issues from an economic perspective, and using economic techniques and thinking, can help conservation scientists meaningfully contribute to real-world policy dialogues.

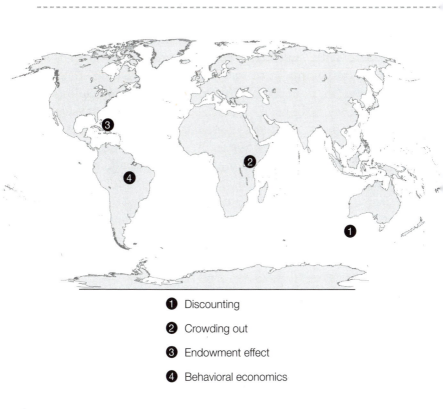

1 Discounting

2 Crowding out

3 Endowment effect

4 Behavioral economics

A Few Wrinkles and Time

How Time, Human Behavior, and Distribution Issues Affect Decision Making

So this it, Chapter 8. Almost there. We hope that in the past seven chapters we demonstrated how important economics is to understanding some of the causes and potential solutions to problems we face in conservation. We explained some of the core concepts in economics that relate to conservation and illustrated some of the ways to integrate economic thinking and analysis into conservation work.

Admittedly, we didn't cover everything, or even part of everything, but rather we cherry-picked topics to try to hit some of the big issues. Opportunity costs, cost-benefit analysis, ecosystem services, valuation, and institutions—these topics underpin some of the fundamental principles and ideas of economics that will be most pressing for conservation. Each one of these topics deserves much fuller treatment, but we didn't go deep on them because (1) that would have been a lot of work and (2) our goal here was simply to introduce you to the core concepts in a field that we think is fundamental to the future success of conservation efforts.

After keeping it simple for seven chapters, though, here we're going to touch on some of the issues that tend to complicate things. We'll look very briefly at some of the operational, behavioral, and ethical issues that affect the integration of economics and conservation.

Operational Issues, or Some Things We Might Have Made Look Simpler than They Really Are

We glossed over a bunch of operational issues in earlier chapters. First, throughout this book we've used cost-benefit analysis as a way to describe an economic approach and to explain some of the key concepts in economics. But as we mentioned in Chapter 3, CBA is by no means flawless. It is difficult to be sure that all the relevant costs and benefits have been accounted for, especially in an environmental decision context. In some cases the downstream effects are not known, and the long-term consequences are sometimes unknowable. So even if we assume that people have the best available information and act rationally (which they often do not), bad decisions can still be made. Despite this, there is plenty of evidence that in general people try to get the most gain with the least cost. So whether one is a stockbroker trading stocks or a farmer trying to graze cattle on common land, gathering all the potential costs and benefits is a good first step in decision making.

The cost-benefit analysis method also makes explicit one of the fundamental issues for conservation: trade-offs. Trade-offs are everywhere; if maintaining Arctic sea ice, stable sea-surface temperatures, or coastal seagrass beds (for polar bears, coral

reefs, and dugong, respectively) came at no cost, then conservation would be a no-brainer. But there are inherent trade-offs when we make decisions to protect ecological systems or degrade them. These are the costs in a typical CBA. Quantifying these trade-offs is sometimes controversial—both monetization and discounting often upset people. That's fine and good. We need a transparent and open debate about how we measure costs and benefits. In some cases it might be impossible to quantify some benefits or apply a discount rate; in others it might be unethical. There is no reason that all costs and benefits need to be monetized or discounted (see Box 5.1). It sometimes helps when they can be monetized so that we can more easily communicate "net" benefits in a common currency to a wider audience. But again, the key is elucidating all the costs and benefits and valuing them in an open and transparent way. We do believe that by doing this we can make better decisions . . . and economic tools provide one way to move us in this direction.

Wouldn't it be great if we lived in a world where scientists did research and then decision makers of all types (e.g., homeowners, governments, businesses) acted directly on the science? However, the institutions we learned about in Chapter 6 are social constructs where decisions are made based on a suite of inputs, not just scientific evidence. So protected areas are not simply placed in areas with the highest species-richness values, but in many cases they are placed where it is politically acceptable (see Box 2.1).

A great lesson on the political aspect of environmental decision making is in Scott Barrett's 2003 book *Environment and Statecraft*. Barrett shows us why the Montreal Protocol, the global agreement limiting the production of ozone-depleting

substances (ODS), was so effective compared to other global environmental agreements. First, the science was bulletproof (science *and* data): "Ahem, there is a gigantic hole in the ozone layer over the South Pole . . . Here's a picture . . . This is immediate bad news for human health." Second, there was a positive cost-benefit outcome. It turned out that the costs of eliminating ODS were way cheaper than the human health costs of a future with an ozone layer full of holes (economics). Third, there was pressure to join (trade benefits and restrictions) once key countries signed up (politics). Fourth, there was social pressure from concerned citizens (power to the people, regardless of the economics).

So while in Chapter 6 we talked about how much sense a payments for ecosystem services (PES) scheme might make for Xavi, we ignored the political and social issues that arise to make such things operational. That topic can fill another field guide someday.

Human Behavior

Let's go back to the assumption that humans are rational. Well, anyone who's ever been in a relationship with another human can probably think of times when they violated this assumption. But economists, remember, have a specific definition of *rational*: people typically making choices that maximize private gain, or utility. There are plenty of situations where we're irrational in this way, too. For example, consider the endowment effect, which holds that we value the things we own more than the exact same thing that we don't own. What?

The concept of the endowment effect comes from a classic experiment from the original boy band of *behavioral economics*—Kahneman, Knetsch, and Thaler (1991). Give a bunch of people a coffee mug. Then tell half of them they own the mug and ask them the minimum price at which they would sell the mug. Tell the other half that they don't own the mug but that they can buy it if they want. How much would they pay for it? In the experiment, the owners would need almost twice as much to part with their mugs as the buyers would pay for theirs. It is just a mug, and the owners only owned it for a few moments . . . and yet the effect is real.

Can we see the endowment effect in conservation? In 1980, a man by the name of Paul Butler became pretty concerned with the likely extinction of the St. Lucia parrot (*Amazona versicolor*). So he led an island-wide campaign that put the image of the parrot onto everything from billboards to stamps to beer mugs. Paul and the folks from the Department of Forestry did radio spots and went to most of the schools on the island. The result: national pride in the St. Lucia parrot, and its future sustained. This was the beginning of the organization Rare. These "pride" campaigns now go on all over the world, protecting spiny lobsters (*Panulirus argus*) in the Bahamas, the Yunnan golden monkey (*Rhinopithecus bieti*) in China, and the red knot (*Calidris canutus*) in Argentina. This is the endowment effect in action. Once the people feel "endowed" with a good, they value it much more and are more willing to fight to protect it.

The rational human is an assumption in many economic theories and models that people will always make decisions that maximize their "net" gain. It's a handy simplifying assumption that often leads to insights. Of course, a deeper understanding of human behavior can strengthen how we understand and manage economic problems, and the whole field is moving this way.

Here's another behavioral complexity—money makes people act weird sometimes. When a social interaction becomes a financial one, it often changes how people behave and feel about it. One impact financial interactions can have is called "crowding out"—as in: financial incentives "crowd out" the altruistic ones. Say you are trying to get your community out to plant some trees at the local school. How do you motivate them to come and help? Well, if you are a science nerd then you'll use this opportunity to test different incentives. How about we pay one group a small sum? We pay another a full day's wage. And we just let the third group think they are doing it out of the goodness of their hearts. Here's what happens. Those offered a small sum actually participate less than those asked to simply volunteer. They are also very grumpy about it. Those offered a full day's wage have high participation rates but are much less satisfied with the day's work than the volunteers doing the same task. Think we just made that up? This little tale comes from the work of John Kerr and friends (2012) in rural Tanzania.

Think about the implications here for perhaps the most popular economic intervention to conserve biodiversity: payments for ecosystem services (Chapter 6). These payments are meant to provide financial incentives for landowners like Isabella to conserve ecosystems instead of converting them for commercial uses. But what if such payments wind up simply crowding out their principled commitment to conservation? Now that it's just a financial transaction, perhaps landowners become less likely to conserve nature for the common good.

Makes sense, doesn't it? We could go on and on about cool

Behavioral economics is the study of how human biases, emotions, irrationalities, and social context affect the economic decisions people make.

examples of behavioral economics, but that's for a future book. For now we'll simply say that it's a burgeoning field that will have a lot to teach us about conservation.

Time and Ethics

None of the authors here claim to be ethicists, although two of us have seen Dr. Seuss's *The Lorax*, and one read a book written by the Dalai Lama. We do imagine, though, that throughout this book your ethical barometer might have been spiking occasionally. For example, when we talked about Xavi's bushmeat hunting and Isabella's decision to log a forest, we only attempted to get at the instrumental values of the forest. We left behind all the cultural, religious, and ethical concerns, like the intrinsic value of the trees themselves. To be clear, all three of us feel that conserving the planet's biodiversity is an end in itself: a moral imperative that doesn't require an economic justification. But economics helps us understand conservation issues and devise solutions.

For example, Srinivasan et al. (2008) used an economic framework to look at the costs and benefits of environmental degradation across the globe for the past 50 years. They found that while most of the benefits of environmental degradation have accrued to the rich countries, 45% of the costs were borne by low-income countries, 52% by middle-income countries, and 3% by rich countries. This inequity certainly raises moral and ethical issues regarding the few rich folks in the world benefiting from the degradation that negatively affects the billions of people marginalized by development. The type of analysis in this paper can help us think about the kind of institutions

> **Discounting** is the practice of changing (typically diminishing) the future values of costs or benefits simply because they occur in the future. For example, $100 is worth more to you today than in five years. A high discount rate (20%) severely reduces the future value of something, whereas a low discount rate (0.5%) puts future values almost on par with current ones.

we might devise to help mitigate such inequity (Chapter 6). We don't cover much on ethics in this book, but a lot of resources are out there. Your starting point should probably be Leopold's 1949 watershed work, *A Sand County Almanac*.

One ethical issue we want to hit upon further relates to the word *time* in the title of this chapter. *Discounting* is important because it is almost always a part of social cost-benefit analyses, and that is the framework on which we've built this book. We mentioned that discounting is a way to bring different time profiles of the costs and benefits of a decision into a single time period. For example, should Taylor take his young kids to see the World Cup in Russia in 2018 or wait until it is in Qatar in 2022? He really wants to go soon, but his kids will appreciate it more later. He's still thinking about how these costs and benefits might change over time.

Like Taylor, individuals typically discount their own private costs or benefits in the future, which seems fine. But the issue gets stickier when we think about social decisions for the longer term, because we enjoy the current benefits and pay the current costs, but any future costs and benefits are borne by others, like our kids and their kids. Especially if a decision (burning coal) produces benefits today (like cheap power) but costs us in the future (like climate change), discounting tends to favor a decision (coal) that is unfair to future generations. That's the ethical problem with discounting.

Even beyond the issue of fobbing off costs on our kids, discounting can raise additional ethical issues. For example, isn't having a healthy Great Bear Rainforest in Canada 100 years from now just as valuable as having it healthy and standing today? Isn't maintaining the diversity of coral reefs in 200 years as worthwhile to a person in the future as it is to us today? Accounting for these future benefits only at a discount is an ethical statement.

Society is more than just a collection of individuals, so the notion that society as a whole will act like a given individual is deeply flawed. Let's say that again. The support for discounting comes from the fact that individuals do discount many future costs and benefits, but that does not mean society should or will discount the future for social issues. For example, the costs of future climate change might not directly affect where you live. You might not live near a coast and therefore not care much about sea-level rise, so as an individual you might think it is a good idea to discount those future costs. But when viewed through the lens of the wider society, those costs might be large and might affect the social welfare of the community or country where you live. So the proper discount rate from the societal perspective might be zero (meaning the future is worth the same as the present), although it might be positive (future is worth less) for you as an individual. Similar concerns exist for the valuation of environmental goods and services; in some cases these items are better considered from a community or societal perspective rather than from an individual viewpoint.

As a result, some folks argue that long-term environmental costs and benefits should simply not be discounted at all. But it's not that easy. A low or zero discount rate could hurt people today for the benefit of folks who aren't even alive yet. Think again of

the coal and climate change example. Given all the uncertainty about the future, should we deliberately choose to raise electric bills on the poor today? It's a tough balance because in many cases (although not all) a high discount rate treats the future environment as unimportant, and a low discount rate might mean a loss for folks today. Tough stuff. This is just the tip of the iceberg when it comes to the ethical and social aspects of discounting.

Where's the Macro?

Throughout the book we focused mainly on microeconomic issues—markets, market failures, price incentives, and individual behaviors, often through the lens of CBA. We only vaguely linked our conservation discussion to macroeconomic issues like fiscal policy (e.g., taxes) and ignored issues like monetary policies (e.g., money supply, interest rates), GDP, and inflation. All of these issues affect our conservation work and the outcomes we aim to achieve. Though as the stand-up economist Dr. Yoram Bauman says, "Microeconomists are wrong about specific things. Macroeconomists are wrong about things in general."[1]

To be sure, several macroeconomic issues are at the heart of ongoing conservation problems—things like economic growth, ecological limits to that growth, and the increasing scarcity of global resources like fresh water and minerals. Most macroeconomic research and policy tends to ignore these issues, which certainly makes it difficult to chart a sustainable future. Incorporating ecological realities into economics is the

---

[1] If you want to make fun of your economist friends, here's your starting point: http://standupeconomist.com/.

purview of ecological economics (Daly and Farley 2011) and, increasingly, environmental economics.

Despite their relevance to conservation, we ignored most of these macroeconomic issues because we wanted to focus on ideas most immediately relevant to specific on-the-ground conservation problems and solutions. The micro view really rests upon the idea that economics is about the study of choices under scarcity and the incentives that affect those choices. Within the world of conservation we certainly face tough choices about what to do with scarce resources. Hence, understanding the incentives and the responses to incentives that people face when making decisions about the environment seems pretty important. We hope we showed you some warts, too, as economic arguments can be flawed and myopic, and they can potentially cross ethical boundaries.

What Do We Do Now?

Our aim in writing this book was to convey an economics approach to decision making in the world of conservation. To get there we described some core economic concepts, created a fictitious land for a case study, and then supported that study with some examples from the real world. We hope that we provided some resources that can help you approach conservation work using the principles of economics, at least as a complement to whatever other approach you are taking. We've mentioned several good examples of the literature throughout (but see also Nelson et al. 2009; Balmford et al. 2011; Polasky et al. 2011; Bateman et al. 2013.) But we also tried to lay out this book in such a way that it can help you in a stepwise fashion.

First, define the issue you are working on and figure out what the economic and non-economic issues are (Chapter 1). Next, try to understand the costs and benefits of the current conditions across all relevant stakeholders (Chapter 2). Then think about how these costs and benefits change under given pressures or as stepwise (marginal) changes (Chapter 3). What specifically are the economic benefits nature is providing (Chapter 4), and how do we value those benefits as well as costs (Chapter 5)? Are there ways in which the values can be captured so that there are "net" societal benefits (chapters 6 and 7)? What are the political, ethical, behavioral, and nonquantifiable issues that need to be made transparent in this work (Chapter 8)?

It's worth repeating: Sometimes it is very difficult to under-stand, let alone quantify, all of the relevant costs and benefits of a conservation issue. We need to be open and honest about this. Sometimes it is too expensive to quantify all of the rele-vant costs and benefits, and sometimes all of the relevant costs and benefits are, well, irrelevant (like when a decision is made for political reasons, corruption, etc.). Despite this difficulty, we only have to look at the literature and the current global de-bate regarding the state of our natural world to conclude that an economic understanding of our conservation and environ-mental challenges will have much to bear on the decisions we make, and the future sustainability of the planet.

OK, we're pretty much done.[2]

---

[2] Because we hate loose ends, we'll tell you that Xavi and Isabella got married. They kept hunting, but they hunted only invasive-pest bushmeat; they restored forest cover; and they made a modest profit on their shade-grown coffee. They donated that profit to a conservation organization—one that employs economists.

But here's a parting note. In his book *Better,* Atul Gawande (2007) says that major future gains in global public health will not come from major advances in cures and vaccines, but rather from improving the delivery of care and interventions we already undertake. We feel the same is true about the future of conservation. It is unlikely to be some fantastic technological or scientific discovery that will improve the fate of the tiger, the Atlantic salmon, and the Amazon in the coming century. Rather, the future of nature depends on people making better decisions and implementing them more effectively and equitably. We'll need economics to achieve this. That's why every conservationist should know at least the basics.

References

Chapter 1

Assunção, J., C. C. e Gandour, and R. Rocha. 2012. Deforestation slowdown in the legal Amazon: Prices or policies? Working paper, Climate Policy Initiative, Rio de Janeiro.

Frank, R. H., and B. S. Bernanke. 2003. *Principles of economics*. New York: McGraw-Hill.

Regalado, Antonio. 2010. Brazil says deforestation rate in the Amazon continues to plunge. *Science* 329:1270–71.

Wilcove, D. S. 2008. *No way home: The decline of the world's great animal migrations*. Washington, DC: Island Press.

Chapter 2

Brashares, J. S., et al. 2004. Bushmeat hunting, wildlife declines, and fish supply in West Africa. *Science* 306(5699):1180–83.

Edwards, D. P., et al. 2011. Degraded lands worth protecting: The biological importance of Southeast Asia's repeatedly logged forests. *Proceedings of the Royal Society* B 278:82–90.

Fa, J. E., et al. 2002. Bushmeat exploitation in tropical forests: An intercontinental comparison. *Conservation Biology* 16(1):232–37.

Fa, J. E., et al. 2009. Linkages between household wealth, bushmeat and other animal protein consumption are not invariant: Evidence from Rio Muni, Equatorial Guinea. *Animal Conservation* 12(6):599–610.

Ferraro, P. J., and L. O. Taylor. 2005. Do economists recognize an opportunity cost when they see one? A dismal performance from the dismal science. *Contributions to Economic Analysis & Policy* 4(1):1–12.

Fisher, B., et al. 2011. Cost-effective conservation: Calculating biodiversity and logging trade-offs in Southeast Asia. *Conservation Letters* 4: 443–450.

Frank, R. H., and B. S. Bernanke. 2003. *Principles of economics.* New York: McGraw-Hill.

Hoekstra, J. M., et al. 2005. Confronting a biome crisis: Global disparities of habitat loss and protection. *Ecology Letters* 8:23–29.

Joppa, L. N., and A. Pfaff. 2009. High and far: Biases in the location of protected areas. *PLoS ONE* 4(12):e8273. doi:8210.1371/journal.pone.0008273.

Keohane, N. O., and S. M. Olmstead. 2007. *Markets and the environment.* (Foundations of Contemporary Environmental Studies series.) Washington, DC: Island Press: xi, 274.

Kumpel, N. F., et al. 2009. Trapper profiles and strategies: Insights into sustainability from hunter behaviour. *Animal Conservation* 12(6):531–39.

Mankiw, N. G. 2003. *Principles of microeconomics.* Cengage South-Western.

Pearce, D. W. 1981. *The dictionary of modern economics.* Cambridge, MA: MIT Press.

———. 1983. *Cost-benefit analysis.* London: Macmillan.

———. 2003. The social cost of carbon and its policy implications. *Oxford Review of Economic Policy* 19(3):362–84.

Rowcliffe, J. M., et al. 2005. Do bushmeat consumers have other fish to fry? *Trends in Ecology & Evolution* 20(6):274–76.

Turner, R. K. 2007. Limits to CBA in UK and European environmental policy: Retrospects and future prospects. *Environmental and Resource Economics* 37:253–69.

Varian, H. 2009. *Intermediate microeconomics: A modern approach.* New York: W. W. Norton & Company.

Chapter 3

Brander, L. M., et al. 2006. The empirics of wetland valuation: A comprehensive summary and a meta-analysis of the literature. *Environmental & Resource Economics* 33(2):223–50.

Coleman Jr., J. L. 1995. The American whale oil industry: A look back at the future of the American petroleum industry. *Natural Resources Research* 4(3):273–88.

Jack, B. K., et al. 2009. A revealed preference approach to estimating supply curves for ecosystem services: Use of auctions to set payments for soil erosion control in Indonesia. *Conservation Biology* 23(2):359–67.

Moser, M., C. Prentice, and S. Frazier. 1996. A global overview of wetland loss and degradation. *Proceedings of the 6th Meeting of the Conference of Contracting Parties of the Ramsar Convention* 10.

Naidoo, R., and W. L. Adamowicz. 2005. Economic benefits of biodiversity conservation exceed costs of conservation at an African rainforest reserve. *Proceedings of the National Academy of Sciences of the United States of America* 102(46):16712–16.

Wilcove, D. S., and L. Y. Chen. 1998. Management costs for endangered species. *Conservation Biology* 12(6):1405–7.

Chapter 4

Balmford, A., et al. 2002. Economic reasons for conserving wild nature. *Science* 297:950–953.

Bateman, I. J., et al. 2011. Economic analysis for ecosystem service assessments. *Environmental & Resource Economics* 48:177–218.

Boyd, J., and S. Banzhaf. 2007. What are ecosystem services? The need for standardized environmental accounting units. *Ecological Economics* 63:616–26.

Daily, G. C., ed. 1997. *Nature's services.* Washington, DC: Island Press.

Daily, G. C., et al. 2000. The value of nature and the nature of value. *Science* 289:395–96.

Fisher, B., R. K. Turner, and P. Morling. 2009. Defining and classifying ecosystem services for decision making. *Ecological Economics* 68:643–53.

Hardin, G. 1968. The tragedy of the commons. *Science* 162:1243–48.

Millennium Ecosystem Assessment. 2005. *Ecosystems and human well-being: General synthesis.* Washington, DC: Island Press.

Tol, R. S. J. 2005. The marginal damage costs of carbon dioxide emissions: An assessment of the uncertainties. *Energy Policy* 33:2064–74.

Chapter 5

Amirnejad, H., et al. 2006. Estimating the existence value of north forests of Iran by using a contingent valuation method. *Ecological Economics* 58:665–75.

Bateman, I. J., et al. 2010. Tigers, markets and palm oil: Market potential for conservation. *Oryx* 44:230–34.

Bockstael, N. E., et al. 2000. On measuring economic values for nature. *Environmental Science and Technology* 34:1384–89.

Boyles, J. G., et al. 2011. Economic importance of bats in agriculture. *Science* 332:41–42.

Costanza, R., et al. 1997. The value of the world's ecosystem services and natural capital. *Nature* 387:253–60.

Das, S., and J. R. Vincent. 2009. Mangroves protected villages and reduced death toll during Indian super cyclone. *Proceedings of the National Academy of Sciences of the United States of America* 106:7357–60.

Driscoll, C. A., et al. 2012. A postulate for tiger recovery: The case of the Caspian tiger. *Journal of Threatened Taxa* 4:2637–43.

Englin, J., J. M. McDonald, and K. Moeltner. 2006. Valuing ancient forest ecosystems: An analysis of backcountry hiking in Jasper National Park. *Ecological Economics* 57:665–78.

Fisher, B., and R. Naidoo. 2011. Concerns about extrapolating right off the bat. *Science* 333:287.

Hope, D., et al. 2003. Socioeconomics drive urban plant diversity. *Proceedings of the National Academy of Sciences of the United States of America* 100:8788–92.

Horton, B., et al. 2003. Evaluating non-user willingness to pay for a large-scale conservation programme in Amazonia: A UK/Italian contingent valuation study. *Environmental Conservation* 30:139–46.

Kramer, R. A., and D. E. Mercer. 1997. Valuing a global environmental good: U.S. residents' willingness to pay to protect tropical rain forests. *Land Economics* 73:196–210.

Luttik, J. 2000. The value of trees, water and open space as reflected by house prices in the Netherlands. *Landscape and Urban Planning* 48:161–67.

Morancho, A. B. 2003. A hedonic valuation of urban green areas. *Landscape and Urban Planning* 66:35–41.

Naidoo, R., and W. L. Adamowicz. 2005. Economic benefits of biodiversity conservation exceed costs of conservation at an African rainforest reserve. *Proceedings of the National Academy of Sciences of the United States of America* 102:16712–16.

Norgaard, R. B., C. Bode, and the Values Reading Group. 1998. Next, the value of God, and other reactions. *Ecological Economics* 25:37–39.

Pearce, D. W. 2001. The economic value of forest ecosystems. *Ecosystem Health* 7:284–96.

Radeloff, V. C., et al. 2011. Housing growth in and near United States protected areas limits their conservation value. *Proceedings of the National Academy of Sciences of the United States of America* 107:940-45.

Richardson, L., and J. Loomis. 2009. The total economic value of threatened, endangered, or rare species: An updated meta-analysis. *Ecological Economics* 68:1535–48.

Ricketts, T. H., et al. 2004. Economic value of tropical forest to coffee production. *Proceedings of the National Academy of Sciences of the United States of America* 101:12579–82.

Thorsnes, P. 2002. The value of a suburban forest preserve: Estimates from sales of vacant residential building lots. *Land Economics* 78:426–41.

Chapter 6

Balmford, A. 2012. *Wild hope : On the front lines of conservation success.* Chicago: University of Chicago Press.

Bateman, I. J., et al. 2010. Tigers, markets and palm oil: Market potential for conservation. *Oryx* 44(2):230–34.

Blackman, A., and J. Rivera. 2011. Producer-level benefits of sustainability certification. *Conservation Biology* 25(6):1176–85.

Burger, N., et al. 2009. In search of effective and viable policies to reduce greenhouse gases. *Environment* 51(3): 8–18.

Ferraro, P. J. 2008. Asymmetric information and contract design for payments for environmental services. *Ecological Economics* 35:810–21.

Ferraro, P. J., et al. 2007. The effectiveness of the US Endangered Species Act: An econometric analysis using matching methods. *Journal of Environmental Economics and Management* 54(3):245–61.

Gardner, T. A., et al. 2013. Biodiversity offsets and the challenge of achieving no net loss. *Conservation Biology* 27(6):1254–64.

Geldmann, J., et al. 2013. Effectiveness of terrestrial protected areas in reducing habitat loss and population declines. *Biological Conservation* 161:230–38.

Li, J., et al. 2011. Rural household income and inequality under the Sloping Land Conversion Program in western China. *Proceedings of the National Academy of Sciences of the United States of America* 108(19):7721–26.

Liu, J. G., et al. 2008. Ecological and socioeconomic effects of China's policies for ecosystem services. *Proceedings of the National Academy of Sciences of the United States of America* 105(28):9477–82.

Naidoo, R., et al. 2011. Namibia's Community-Based Natural Resource Management program: An unrecognized payments for environmental services scheme. *Environmental Conservation* 38(4):445–53.

Ostrom, E. 1990. *Governing the commons : The evolution of institutions for collective action.* Cambridge, UK: Cambridge University Press.

Pretty, J. 2003. Social capital and the collective management of resources. *Science* 302:1912–14.

Roman, J., et al. 2013. The Marine Mammal Protection Act at 40: Status, recovery, and future of US marine mammals. *Year in Ecology and Conservation Biology* 1286:29–49.

Wunder, S. 2007. The efficiency of payments for environmental services in tropical conservation. *Conservation Biology* 21(1):48–58.

Chapter 7

Adams, V. M., R. L. Pressey, and R. Naidoo. 2010. Opportunity costs: Who really pays for conservation? *Biological Conservation* 143:439–48.

Agarwala, M., et al. 2010. Paying for wolves in Solapur, India and Wisconsin, USA: Comparing compensation rules and practice to understand the goals and politics of wolf conservation. *Biological Conservation* 143:2945–55.

Ban, N. C., and C. J. Klein. 2009. Spatial socioeconomic data as a cost in systematic marine conservation planning. *Conservation Letters* 2:206–15.

Bateman, I. J., et al. 2010. Tigers, markets and palm oil: Market potential for conservation. *Oryx* 44:230–34.

Dickman, A. J., E. A. Macdonald, and D. W. Macdonald. 2011. A review of financial instruments to pay for predator conservation and encourage human-carnivore coexistence. *Proceedings of the National Academy of Sciences of the United States of America* 108:13937–44. doi: 10.1073/pnas.1012972108.

Dinerstein, E., et al. 2012. Enhancing conservation, ecosystem services, and local livelihoods through a wildlife premium mechanism. *Conservation Biology* 27(1):14–23. doi: 10.1111/j.1523-1739.2012.01959.x.

Environmental Investigation Agency. 2013. *Hidden in plain sight: China's clandestine tiger trade.* London: EIA.

Ferraro, P. J., and A. Kiss. 2002. Direct payments to conserve biodiversity. *Science* 298:1718–19.

Gardner, T. A., et al. 2012. A framework for integrating biodiversity concerns into national REDD+ programmes. *Biological Conservation* 154:61–71.

Hargrave, J., and K. Kis-Katos. 2013. Economic causes of deforestation in the Brazilian Amazon: A panel data analysis for the 2000s. *Environmental and Resource Economics* 54:471–94.

Horton, B., et al. 2003. Evaluating non-user willingness to pay for a large-scale conservation programme in Amazonia: A UK/Italian contingent valuation study. *Environmental Conservation* 30:139–46.

Kirkby, C. A., et al. 2010. The market triumph of ecotourism: An economic investigation of the private and social benefits of competing land uses in the Peruvian Amazon. *PLoS ONE* 5:e13015. DOI: 10.1371/journal.pone.0013015.

Koh, L. P., and D. S. Wilcove. 2007. Cashing in palm oil for conservation. *Nature* 448:993–94.

Moro, M., et al. 2013. An investigation using the choice experiment method into options for reducing illegal bushmeat hunting in western Serengeti. *Conservation Letters* 6:37–45.

Morse-Jones, S., et al. 2012. Stated preferences for tropical wildlife conservation amongst distance beneficiaries: Charisma, endemism, scope and substitution effects. *Ecological Economics* 78:9–18.

Naughton-Treves, L., R. Grossberg, and A. Treves. 2003. Paying for tolerance: Rural citizens' attitudes toward wolf depredation and compensation. *Conservation Biology* 17:1500–11.

Strassburg, B. B. N., A. Kelly, A. Balmford, R. G. Davies, H. K. Gibbs, A. Lovett, L. Miles, C. D. L. Orme, J. Price, R. K. Turner, and A. S. L. Rodrigues. 2010. Global congruence of carbon storage and biodiversity in terrestrial ecosystems. *Conservation Letters* 3: 98–105.

West, P., J. Igoe, and D. Brockington. 2006. Parks and peoples: The social impact of protected areas. *Annual Review of Anthropology* 35:251–77.

Wikramanayake, E., et al. 2011. A landscape-based conservation strategy to double the wild tiger population. *Conservation Letters* 4(3):219–27.

Chapter 8

Balmford, A., et al. 2011. Bringing ecosystem services into the real world: An operational framework for assessing the economic consequences of losing wild nature. *Environmental and Resource Economics* 48:161–175.

Barrett, S. 2003. *Environment and statecraft: The strategy of environmental treaty-making.* Oxford, UK: Oxford University Press.

Bateman, I. J., et al. 2013. Bringing ecosystem services into economic decision-making: Land use in the United Kingdom. *Science* 341(6141):45–50.

Daly, H. E., and J. Farley. 2011. *Ecological economics: Principles and applications.* 2d ed. Washington, DC: Island Press.

Gawande, A. 2007. *Better: A surgeon's notes on performance.* New York: Metropolitan.

Kahneman, D., J. L. Knetsch, and R. Thaler. 1991. Anomalies: The endowment effect, loss aversion, and status quo bias. *Journal of Economic Perspectives* 5(1):193–206.

Kerr, J., et al. 2012. Prosocial behavior and incentives: Evidence from field experiments in rural Mexico and Tanzania. *Ecological Economics* 73(15):220–27.

Leopold, A. 1949. *A Sand County almanac, and sketches here and there.* New York: Oxford University Press.

Nelson, E., et al. 2009. Modeling multiple ecosystem services, biodiversity conservation, commodity production, and tradeoffs at landscape scales. *Frontiers in Ecology and the Environment* 7(1):4–11.

Polasky, S., et al. 2011. The impact of land-use changes on ecosystem services, biodiversity, and returns to landowners: A case study in the state of Minnesota. *Environmental and Resource Economics* 48:219–242.

Srinivasan, U. T., et al. 2008. The debt of nations and the distribution of ecological impacts from human activities. *Proceedings of the National Academy of Sciences of the United States of America* 105(5):1768–73.

Index